Praise for
The Jasons by Ann Finkbeiner

"Never heard of the Jasons? These crack U.S. scientists with top-secret clearance have advised the Pentagon since 1960 on some of its toughest problems. Finkbeiner's unusual access makes for a true story that reads like a Tom Clancy novel."　　　　　　　　　　　—*Wired*

"A glimpse into the minds of the scientists who have taken on the job of advising the government of the world's most powerful nation."
　　　　　　　　　　　　　　　　　　　　　　　　　　　—*Discover*

"A fascinating tale of policy, money, the value of independent advice, and the seduction of science. [The Jasons]: independent check on Pentagon hubris or naive, arrogant enabling tool of the military-industrial complex?"　　　　　　　　　　—*The Seattle Times*

"An important investigation into the relationship between science and government. At heart, *The Jasons* is a meditation on morality."
　　　　　　　　　　　　　　　　　　　　　　　　　　　　—*Seed*

"Excellent. A fascinating, often moving, account . . . of how this small cadre of brilliant American scientists provided vital technological support to our government over a period of forty-six years following the end of World War II."　　　　　　　—*The New York Sun*

"Insightful. An illuminating history . . . provides an overview of a group that has existed in the shadows since the 1960s to give the Pentagon—and other government agencies—independent advice."
　　　　　　　　　　　　　　　—*Aviation Week & Space Technology*

T0176306

PENGUIN BOOKS

THE JASONS

Ann Finkbeiner is a freelance science writer who writes about cosmology and who runs the graduate program in science writing at Johns Hopkins University. She lives in Baltimore with her husband, a retired physicist.

The Jasons

THE SECRET HISTORY
OF SCIENCE'S
POSTWAR ELITE

Ann Finkbeiner

PENGUIN BOOKS

PENGUIN BOOKS

Published by the Penguin Group
Penguin Group (USA) Inc., 375 Hudson Street, New York, New York 10014, U.S.A.
Penguin Group (Canada), 90 Eglinton Avenue East, Suite 700, Toronto,
Ontario, Canada M4P 2Y3 (a division of Pearson Penguin Canada Inc.)
Penguin Books Ltd, 80 Strand, London WC2R 0RL, England
Penguin Ireland, 25 St Stephen's Green, Dublin 2, Ireland (a division of Penguin Books Ltd)
Penguin Group (Australia), 250 Camberwell Road, Camberwell,
Victoria 3124, Australia (a division of Pearson Australia Group Pty Ltd)
Penguin Books India Pvt Ltd, 11 Community Centre, Panchsheel Park, New Delhi – 110 017, India
Penguin Group (NZ), cnr Airborne and Rosedale Roads, Albany,
Auckland 1311, New Zealand (a division of Pearson New Zealand Ltd)
Penguin Books (South Africa) (Pty) Ltd, 24 Sturdee Avenue,
Rosebank, Johannesburg 2196, South Africa

Penguin Books Ltd, Registered Offices:
80 Strand, London WC2R 0RL, England

First published in the United States of America by Viking Penguin,
a member of Penguin Group (USA) Inc. 2006
Published in Penguin Books 2007

3 5 7 9 10 8 6 4

Copyright © Ann Finkbeiner, 2006
All rights reserved

THE LIBRARY OF CONGRESS HAS CATALOGED THE HARDCOVER EDITION AS FOLLOWS:
Finkbeiner, Ann K.—.
The jasons: the secret history of science's postwar elite / Ann Finkbeiner
p. cm.
Includes index.
ISBN 0-670-03489-4 (hc.)
ISBN 978-0-14-303847-4 (pbk.)
1. JASON Defense Advisory Group—Biography. 2. Scientists—United States—Biography.
3. Physicists—United States—Biography.
4. Group work in research—United States—History—20th century.
5. Science—Research—United States—History—20th century. I. Title
Q141.F536 2006
509'2273—dc22
[B] 2005043471

Printed in the United States of America
Set in Minion
Designed by Francesca Belanger

Except in the United States of America, this book is sold subject to the condition that it shall not, by way of trade or otherwise, be lent, resold, hired out, or otherwise circulated without the publisher's prior consent in any form of binding or cover other than that in which it is published and without a similar condition including this condition being imposed on the subsequent purchaser.

The scanning, uploading and distribution of this book via the Internet or via any other means without the permission of the publisher is illegal and punishable by law. Please purchase only authorized electronic editions, and do not participate in or encourage electronic piracy of copyrighted materials. Your support of the author's rights is appreciated.

for James Calvin Walker

Contents

Introduction

Jasons

"Jason is a very effective thing, it's unique, it was a good idea that worked. And the story should be told so that it will happen again somewhere."

—Richard Muller, interview, June 2002

I first heard of Jason in 1990. I went with my husband to a dinner given by the Johns Hopkins University physics department for an honored guest. The guest, Freeman Dyson, was being honored for several reasons: early on he had refined the theory of how the world behaves at its most fundamental; he'd spent his career since at the prestigious Institute for Advanced Study; he wrote books and elegant, unpredictable essays; and according to the other physicists, he was brilliant.

He was certainly brilliant at the dinner—singular, charming, and a superb storyteller. One story was about the night he spent on the Mexican border looking for the infrared signatures of drug runners. The physicists at the dinner were interested in the infrared signatures. I was interested in Dyson, who looked like an aged bird, crawling around on the Mexican border at midnight. "Why would a physicist be looking for drugs?" I asked. He was doing a Jason study, giving the government technical advice, he said, thinking about clever ways to detect drug smugglers, "and we were talking to the border patrol people. They asked us if we'd like to see them in action. It was very educational."

"And what's Jason?" I asked. I don't remember what he said, but it wasn't an answer. So after dinner on the way home, I asked my husband, "What's Jason?" He said he didn't really know, but it was sort of a top-secret bunch of academic physicists who advise the Defense Department. "I think John Wheeler was one of them," he said, and digressed into a story about Princeton in the late 1950s, when he was a graduate student and when John Archibald Wheeler was faculty and had already explained nuclear fission but hadn't yet named black holes. Anyway, my husband had applied for a federal fellowship but would be turned down unless Wheeler got around to writing a recommendation. Wheeler had felt so bad about this situation that he said, "Never mind. I'll just go down to Washington and recommend you in person," and shortly thereafter my husband got the fellowship.

I believed this story. I had interviewed Wheeler a few years before, and he'd seemed unusually familiar, for a scientist, with the ways of Washington. The following week I asked my husband's physicist-colleagues "What's Jason?" and collected the following: *Jason* is an acronym for July-August-September-October-November, the months this group of academic physicists met secretly to solve the problems the Defense Department couldn't. The Jasons were the crème de la crème. "Yes, Wheeler was a Jason, and I know others," said one physicist. "Want me to call them for you?"

These stories I didn't believe. An acronym for months sounded silly. Academics couldn't meet that long during the school year because they had institutional responsibilities for classes, visitors, departments, and committees; and they wouldn't meet during their precious summers, which were free for research. Anyway, if the Jasons were such cream, they wouldn't work for the Defense Department. In the first place, the Defense Department wouldn't need them. Any physicists the department needed—because radar, sonar, the trajectories of missiles, and the explosions of bombs are just physics—were in its own laboratories or in the government's national laboratories.

In the second place, the cream wouldn't necessarily be in defense or government labs, where research was determined by political program. The cream would be in the universities, in the academy; the cream of the cream were tenured professors at the big research universities, at Princeton, MIT, Stanford, Harvard, Berkeley, and Caltech.

Nor would the cream necessarily be doing defense research. Defense physics is generally applied research; it solves problems by applying other physicists' more basic work. Its result is usually not knowledge but technology. The adjective *applied*, when used by physicists, is not a compliment. Academic physics is pure research; its direction is determined by what academics are curious about and can get funding for. It is, of all the sciences, the most fundamental; its questions are about matter's basic nature and the forces that govern the known universe. Moreover, pure research is considered innocent, neither moral nor immoral. Applied research, whose technologies have potential for harm, comes accompanied by difficult moral decisions.

Academic physicists have secure jobs, roomy salaries, interesting students, the respect of the physics community, and at most twenty productive, creative years. They wouldn't need a thing the Defense Department had to offer. They wouldn't take the time or trouble to be Jasons.

So I took up my husband's colleague on his offer to make phone calls and found that, though I might be generally right about academic physicists, I was mostly wrong about Jason. Jason had been formed in 1960, partly through Wheeler's Washington connections; Dyson was an early member. Jason is asked questions primarily by the Defense Department but also by other government agencies. Between one-half and three-quarters of their studies are classified. They meet in a building leased from General Atomics in La Jolla, California. They work together for six or seven weeks from mid-June through the end of July. Depending on how they're counted, they number between thirty and sixty at any given time; all have top secret clearances. I was right about the name: it has nothing to do with the months they meet and it isn't even an acronym; it's just a name, the Greek myth of Jason and the Argonauts. *Jason* is both a collective and a proper noun: if you belong to Jason, you are a Jason. A few people use the term *Jasonite*, like some biblical tribe or door-to-door religious salesman.

And they are indeed the cream. They are either in the country's best science departments or were trained there. Of the roughly one hundred Jasons over time, forty-three have been elected to the National Academy of Sciences; eight have won MacArthur awards; one won mathematics' Fields Medal; eleven have won Nobel Prizes. Wheeler wasn't actually a Jason but was one of their first advisers, as were Edward Teller, Eugene Wigner, and Hans Bethe; Wigner and Bethe also won Nobel Prizes.

I read everything I could find, published and not published, about Jason. I couldn't find much: Jason is allergic to being public. I did find a remarkable archive of transcribed interviews of Jasons done in the mid-1980s by Finn Aaserud, a historian of physics who is now director of the Niels Bohr Archive in Copenhagen, on Jason's early days. I learned that Jasons have never had a real leader; what they call "Jason management" is a rotating chair and a steering committee that together run Jason like an academic administration, set up primarily to take care of logistics so that faculty can get on with their research.

I also learned that from the beginning Jason was set up to be independent. Jasons were and still are invited to join not by the government, but by other Jasons. Jason management, working together with the government sponsor, decides which questions to study. Jasons themselves work only on those studies that they want to work on; they show up for summer studies when and how often they want to. They're government advisers with satisfying, full-time jobs elsewhere; a Jason can be fired at any minute and not particularly care.

What I couldn't find was the reason academic physicists would turn from studying ultimate reality to, say, shooting down missiles. So for the next two years I chased all over the country interviewing Jasons.

I can't tell you much about this one particular Jason except that he looks like an unusually intelligent child; he looks chronically delighted and wary. I walked into his office, and the first thing he did was lay down the ground rules that Jason management gave him for talking to me. "If you want to talk to her," they told him, "that's fine. You certainly can't go into anything classified, and it should not be up to you to name other Jasons. If you want your own name revealed, that's up to you." This Jason didn't want his name revealed. "I can be referred to as a biochemist Jason," he said, "because I think you'd want to communicate the idea that not all Jasons are physicists and you are now interviewing a Jason that is not and never has been a physicist. So I'm a chemist, more of a biologist hybrid, a chemical biologist. Otherwise I'm anonymous. And I won't be naming names of other Jasons.

"So," he began, "my name is Dr. X"—he thought this was funny—"and I am a Jason, and I'm happy to talk with you about Jason." The first time Dr. X met the Jasons, in 1994, he'd been asked as an outside expert to give them a briefing for a study they were to do. He'd heard of Jason before, he said, because "I guess it was in the seventies the Jasons got outed." But meeting them for the first time was "just the most amazing experience," he said. "The line that kept going in my head was 'Who are these guys?' "

Dr. X arrived at General Atomics, which, he said, "is kind of quaint. The building looks very sixties, and it hasn't been remodeled forever, money is not being spent." He went through a security check with the guard at the outer gate, another check with the guard who buzzed him through the door, and another with the guard who looked at his badge. He looked down a hallway, toward what he thought of as the inner sanctum, and thought again, "Who are these guys?"

He waited in the briefing room for an announcement of his talk's classification level, then began his presentation, which was about a new way of using molecules to do computing. "I went in very pragmatic—which I think is exactly what the Jasons want—saying, 'Okay, we're going to use molecules to do computing. How many molecules would we need? How would we make them?' I planned I would go through my slides, give my talk."

This is the way scientists usually give talks, not by reading from prepared papers but by showing slides in order, discussing each slide, and taking questions at the end. Jasons didn't wait until the end; they interrupted immediately. "The first slide was up there for fifteen minutes. So I just kept talking, going with the interruptions. But the interruptions were fascinating. And they'd ask just the right question, or even the wrong question but the answer would show that it's the wrong question and they'd say, 'Oh, oh no, that isn't it. What I mean is—.' So I'm taking them through it, step by step. And they just were just"—and here Dr. X made a bathtub drain sound—"sucking in everything I could give them. And they're not only evaluating what I was talking about, but in real time they're designing a computational approach to do in silicon what I was describing would be done in molecules. And they came to the conclusion, in real time, that the silicon would win. And I was just so impressed." Dr. X thought to himself again, "Who *are* these guys?"

In the beginning all Jasons were physicists; even now the majority are. Physicists as a group are off-scale intelligent, gossipy, competitive, relentlessly rational, and promiscuously curious. Their taste for pure research comes from the belief that some given truth about the real world exists and that they can find it; and when they've found it, sooner or later they'll all agree it's the truth. They are fascinated by the congruence between their theories and reality. They are famous for arrogance: because physics is the most fundamental science, physicists say, a physicist can solve any problem in any field by figuring it out from first principles, from the ground up. To a physicist, arrogance is the happy result of having effective tools, enough brains, a lively outlook, and a carefully defined subject. For a physicist, being called "arrogant" is not necessarily a criticism; for, say, a biologist, it might be. I told a government biologist who is one of Jason's sponsors that I wanted to call this book *The Arrogance of Physicists,* but my editor thought it wouldn't be nice. The biologist laughed and said, "The Jasons would like it."

Physicists have also traditionally been the government's science advisers, a tradition that began with their central role in World War II's Manhattan Project. Manhattan Project physicists took their pure research into atomic physics and applied it to an atomic bomb, the prototype of a harmful technology forcing a moral decision. The physicists hadn't necessarily felt guilt for building their bomb—they knew Hitler's physicists were trying to build one, too—but they did feel an acute sense of public responsibility. With some notable exceptions, physicists spent the following decade lobbying the government to move control of nuclear research out of military hands and into civilian hands; to stop the proliferation of nuclear weapons; and to ban the tests of new ones.

Then in 1957, when a newly intercontinental Soviet missile carried the Soviet Sputnik—which could as easily have been a nuclear bomb—across American skies, physicists broadened their interests to include missile defense. The country went into shock over Soviet technical superiority; the government once again found a use for scientists and created a number of venues for getting their advice. Some ex–Manhattan Project physicists proposed the formation of a group that could, according to John Wheeler, "inject new ideas into national defense." Around two years later, in 1960, Jason went into the business of advising the government. The first Jasons were young Manhattan Project physicists or their students.

Given Jason's heritage, its subsequent history seems inevitable. Scientists can't become governmental science advisers without learning to operate in the world of politics. In the world of science, rules are based on logic: the truth, the one right answer on which everyone agrees, is found by publicly sorting through and understanding the disagreements; suppressing disagreement only slows progress. In politics, however, rules are based on human nature: no single answer will ever be right for all people, but decisions have to be made anyway; decisions are made by working through colleagues, or a party, or an administration—that is, through a network of loyalties; loyalty can decree that in the interest of getting things done, disagreement—at least in public—be suppressed. Science advisers are most effective when they operate on political rules, like political insiders. Jasons explicitly and repeatedly insist, however, that their whole value lies in operating like logical, outspoken, unbeholden, independent outsiders.

Independent outsiders are useful to the government but aren't always welcome: they feel free to disagree publicly; their reasons for giving advice are often different from the government's reasons for requesting it; and they are occasionally politically naïve and usually unpredictable and uncontrollable. "Anytime you invite a team as bright as Jason to be devil's advocates on a program you're committed to," said one sponsor, an ex–secretary of defense, "you're asking for trouble." So once you've answered the question of who these Jasons are, the next question is how they have managed to last so long.

During Jason's early years, its sponsor was a small, unconventional entity in the Department of Defense appropriately named the Advanced Research Projects Agency, or ARPA. Jason's studies tracked not only its own postwar sense of responsibility but also ARPA's scientific missions of defense against ballistic missiles (for example, whether missiles can be shot down by lasers) and verification of cheating on nuclear test bans (for example, whether nuclear explosions might be hidden in underground caverns). ARPA asked interesting questions; Jason had relevant answers; and ARPA in turn listened. Jason flourished, riding what are now called the golden years of science advising.

Then in 1966 Jason took the sense of responsibility it had learned from the Manhattan Project and applied it to the escalating war in Vietnam. Jasons did a series of studies, one of which, a barrier of sensors, was meant to damp down the war. But the Defense Department uses technology for its own purposes, which are not necessarily those that scientists intend; and the well-intentioned sensor barrier was turned into the lethal electronic battlefield. The electronic battlefield was not in the same league as nuclear weapons, but Jasons were unhappy about their part in its creation. A few years later a leaked report from the Defense Department, published as *The Pentagon Papers*, detailed the Jason studies and, as Dr. X said, "outed" the Jasons. Back home at their virulently antiwar universities, the Jasons caught hell: they were picketed, called names, shouted down, and threatened physically. For Jasons, being publicly attacked for a war about which they were already miserable was deeply disturbing, and a number of them quit.

After Vietnam the glory years were over and the considerable influence that the independent science advisers had had in the government began to fade. At the same time ARPA was increasingly unhappy with the Jason physicists: the agency's own problems had been changing with the scientific times and now included sciences other than physics. So Jason adapted. It tailored its membership to ARPA's needs and added more nonphysicists. It also broke ARPA's sponsorship monopoly and added new sponsors—the intelligence community, the navy, and the Department of Energy.

Jason's changing sponsors meant that Jason itself had to change. Because new sponsors, particularly the navy, had differing requirements for clearances, Jasons could no longer talk to all other Jasons about all studies. Because the Department of Energy's problems were not necessarily military, Jason for the first time did studies unrelated to defense. Unclassified studies on climate echoed the country's growing environmental movement, were intellectually interesting to these constitutionally curious physicists, and were morally comforting to the war-shocked Jasons. Accordingly, Jason added even more nonphysicists to its membership. Different Jasons began specializing in studies for different sponsors, and Jason began fragmenting into clusters.

Meanwhile, throughout the Cold War, Jasons continued work for what was now called DARPA—for Defense Advanced Research Projects Agency—on the perennial problems of missile defense. The results were mixed. Some—like a study helping DARPA overcome the distortion that a turbulent atmosphere inflicts on the images of enemy satellites—were classic matches between government and science. Others—like Jason's proposal to base air force missiles on small mobile navy submarines—were case studies in how political realities can wipe out perfectly good technical ideas.

With the end of the Cold War, Jason undertook a long series of studies on the health of the nation's nuclear stockpile. The studies coincided with individual Jasons' missions of banning further tests of nuclear weapons. They also implied national policy and therefore forced Jason to define the boundary between technical and policy advice. Jasons had to decide whether to give technical advice and stay away from policy, or whether their technical expertise gave them a unique perspective on certain policy problems. They had to decide whether to go public with the results of their studies: was testifying to Congress or talking to the newspapers a betrayal of their sponsors' trust, or was speaking publicly simply the right of all citizens—and perhaps an obligation for those with technical expertise?

Over time Jason's management of these and other insider-outsider issues changed the question from how Jason got along with its sponsors to whether Jason could prevent its own dissolution. What had been a cohesive group of physicists, having the same habits of thought and speaking the same scientific language, turned into a group that includes mathematicians, engineers, computer scientists, geoscientists, oceanographers, chemists, and finally the ultimate nonphysicists, biologists and biologist-hybrids like Dr. X.

Dr. X was invited to join Jason in 1996, about two years after his talk as an outside expert. In his first summer at Jason, he went to a briefing given by the Department of Energy on aging. As a biologist, he was interested in aging and at first thought, "Boy, Department of Energy's in on this?" Then he thought, "Well, you know, the Department of Energy has the Genome Project. Maybe they're looking for gene clusters that affect aging." But then the briefing's level of classification was announced, and he couldn't understand why it was so high, and he wondered whether it meant he might learn some things "that are really cool that aren't in the

outside scientific world." He thought, "So have they really got secrets about aging?" He noticed that the other Jasons at the meeting
were "very much the hard-core physics crowd," which was odd, but
he thought, "Well, good for them." Then the briefer put the first
slide up on the screen, announcing a talk on the aging not of people but of nuclear warheads. "And of course any idiot would know
that," he said. "I stayed until the break." Afterward he told the hard-
core physicists about his mistake, and they thought it was funny—
"this poor nonphysicist thinks that aging is aging."

Jasons say their scientific diversity has been their biggest change.
The next biggest change is social and has had an unexpected impact. Back in 1960, when academia was leisurely and wives—Jasons
were then all men—didn't go to work, Jasons could move to California for six weeks in the summers without jeopardizing jobs or
families. And in fact the wives became friends, the kids grew up together, and Jason became its own family. But the new, scientifically
diverse Jasons have more calls on their time and less flexible sched-
ules, and new Jasons' wives, and now a few husbands, have jobs to
keep. As a result, Jasons increasingly show up only for disjointed
fractions of those summers. One worry is, of course, that Jason will
lose the cohesiveness and intensity that sponsors find useful. The
other worry is that Jason will turn into just one more panel of scientists having two-day meetings in which they read reports written
by other people and then write their own—not the sort of thing for
which a hotshot scientist would give up summers. No one's quite
sure what to do about it, or even whether they'll stay in business.
"We've got a bunch of new Jasons, and they're just terrific," said the
current chair, Roy Schwitters, a physicist at the University of Texas
at Austin. "We just don't know—we call it sticking—whether they'll
stick or not."

Does it matter? Jasons think they've done the country good, but
they haven't got much clear evidence. Jason doesn't and never has

had a formal way of keeping track of whether its recommendations were taken and if so whether they worked. When asked, Jasons shrug and explain that certain Jasons keep track of certain sponsors and therefore have a sense of the outcome of those sponsors' studies. And the government insiders, when asked, say that in any case political decisions have so many aspects and balances and parties, no one can track the influence of any one recommendation.

Jasons say they've hit no home runs—though they're being a little modest since, aside from their inadvertent invention of the electronic battlefield, their DARPA study about undistorting the atmosphere helped change astronomy, and one of their stockpile studies allowed a president to sign a test ban treaty. They say they're best at shooting down scientifically stupid ideas and that they've saved the government millions if not billions of dollars. They also say, as do the government insiders, that whether Jason matters depends less on Jason than on the government: in the political world, personal relationships count; and without these relationships, scientists can advise all they want to but no one will listen.

In fact for some time now successive administrations have listened less and less to the advice of independent scientists. The advisory apparatus has been moved farther and farther away from the government's center, and decisions about partly technical issues like missile defense and nuclear safety have become mostly political. In 2002 Jason's relationship with DARPA broke up over Jason's independence: when DARPA's director told Jason to accept certain members, Jason refused and DARPA fired Jason.

But the Defense Department had second thoughts, and Jason got itself rehired by DARPA's boss, the director of defense research and engineering. For the government, Jason is unique. No other science advisers of such high caliber have worked together in so close a group for so long a period, doing their own calculations. Few other science advisers can study such a wide range of subjects. Few others know so much about their sponsors' needs yet gain so little from giving them advice. And Jason is cost effective. Jasons are now paid around $850 per day—a tidy sum but a mere fraction of what they could make consulting for industry—and Jason's budget is $3.5 million. "I put it in the top one percent of Defense Department expenditures," said an ex-director of DARPA. "I mean, what's a few million dollars? There's no question about that."

For this book I talked to thirty-six Jasons, roughly half of the membership, active and semiretired senior. They were, every one of them, interesting. Most were socially easy. Some were in love with their own brainy selves, delighted to share with me their talents for speaking at length on any subject without taking a breath or noting a response. Many were funny. Some were so individual, so idiosyncratic, I'd never met their likes before.

On the whole, they seemed to have the qualities that make for good advisers. Those who were arrogant-as-charged were annoying, but they didn't seem to let their arrogance get in the way of having open minds, or admitting they were wrong, or being genuinely enchanted with something new. Some were so much at ease in both the political and the scientific worlds, I didn't see how they could contain the contradictions. A few radiated unimpeachable integrity. All were so deeply independent that when I told one Jason I worried that they'd go unilateral on me and collectively refuse to talk, he said, "No, that's not the way things are done," meaning that not even Jason dictated what Jasons do.

I've been interviewing scientists for twenty years, and though the Jasons were as generous and graceful as scientists usually are, these interviews were exceptions in two ways. One was that the Jasons were unusually interventionist interviewees. They were interested in my slightly unreliable tape recorder, which has an automatic reverse:

they'd interrupt to tell me my machine had clicked; or to ask if I wanted to change to side B; or to explain why low batteries messed up the recording; or to suggest that I buy a higher-tech digital recorder—"The dumb thing has got a tape, hasn't it? Yes, well, all right. There you are." My tape recorder could also be controlled by its little lapel microphone: one Jason switched control of the recorder to the microphone, then put the microphone out of my reach on his lapel; a politician might do this sort of thing, but a scientist never does. Jasons asked who was publishing the book and whether I already had a contract and what I thought my approach would be and what I saw as the book's point. Several Jasons thought the subject would be unsustainable or uninteresting—"I honestly don't know whether there's a book on this subject." Six Jasons came up with eight different storylines, all reasonable. One recommended that the book be balanced and not make Jasons out to "be either giants or jerks." A lot of Jasons predicted cheerfully that no one would buy it.

The other exception was their secrecy. Secrecy is antiscience. When scientists discover something, they publish their data, methods, and conclusions in the open literature, where their competitors can read it; the competitors point out every possible weakness, then propose improvements whose weaknesses are, in turn, pointed out by the original scientists, who then propose the next round of improvements, and so on far into the night. By this process scientists decide which pieces of knowledge are believable, then use those pieces as the basis for further knowledge. The entire structure thus depends on openness.

Because openness is in the air that scientists breathe, you usually can't shut them up. They explain, then explain another way, digress, compare, generalize, give examples, give counterexamples, offer opinions, offer opposing opinions, then explain once more, and they rarely refuse requests for interviews. Of the forty-six

Jasons I asked to interview, ten either never replied or said no—
"I am flattered by your request but would prefer to decline." I got
e-mail from the Jason management saying, "We, frankly, could not
identify an up-side for our organization but could identify potential down-sides to such a book. We wish you all the best in your future endeavors." Several talked to me because, as one said, "you are
going to write a book and I would rather you write an accurate
book than one based on limited information because nobody
would talk to you." I told one Jason that I knew I was a mixed bless-
ing for them, and he said, "Yes, but more like having to make the
best of a bad deal." And I was advised indirectly not to try any interviewer's tricks: "We've talked about you many times," said Dr. X,
and he wasn't alone in saying this. "And I think it's safe for you to
assume that anything you say to any Jason is going to be reported
back to the other Jasons. Because so far it has."

They were reticent not because they worried that they'd give
away classified information: if they couldn't remember whether
something was classified, or whether some part of it was classified,
they didn't talk about it at all. They were reticent because they had
had to learn insider rules. "Our sponsors trust us with relatively
sensitive information," said one Jason—"relatively sensitive" meaning information that, if not classified, was still guarded because it
was part of the sponsor's territory, or because if revealed it would
make the sponsor look less than brilliant or because it was just
work in progress—"and we don't want to see what we do paraded
in the headline in *The New York Times* or anywhere else." The
ground rules Dr. X outlined to me included the stipulation that Jasons should "respect the sponsor's desire for anonymity with respect
to particular studies."

For some Jasons, reticence was clearly a strain on their scientific
nerves. A few handled this problem by talking not-for-attribution.
Some talked in generalities that were to the point but gave away

nothing I might not easily know. Some—Dr. X was one—managed the interview: they had several stories that ought to keep me happy and away from subjects they wanted to avoid; they had decided how much they thought I should know and refused politely to go farther. Some handled it by giving limited, strictly relevant answers, no examples, no digressions—if I wanted the answer, I had to ask the question:

"How did you manage to observe the siege at Khe Sanh?"

"People were being helicoptered in and out."

"And you went to Khe Sanh as—?"

"An adviser."

"As a member of Jason?"

"Actually it was in association with another agency that was helping."

"And you went there to find out if they were putting the sensors in correctly?"

"That's right."

"And they were?"

"Yup."

A few who talked more easily worried out loud. "I'm probably being too responsive," said one. "Compared with your other informants, am I singing like a bird?"

Though the Jasons' reluctance to speak might have been an effort to protect classified material and the trust of the sponsor, the ground rules about revealing names were not. They were a way to protect Jasons who didn't want to be known as Jasons. For many of these anonymous Jasons, the reason was historical, what one Jason called their "racial memory" of the Vietnam War and the ensuing personal attacks; thirty years later I still couldn't get an official list of Jason members. Dr. X's reason for anonymity, however, was current. Being a Jason, he said, made him nervous. He said that other young Jasons felt the same. Some Jasons, reluctantly and briefly, referred to

xxviii   Introduction

the feeling as worrisome knowledge or loss of innocence. One told a story about a Nobel Prize–winning Jason at a Strangelovean defense briefing during the Cold War, hearing for the first time about computer simulations of massive nuclear exchanges between the United States and the USSR and the deaths of twenty million people, and yelling from the back of the room, "Helllppp!"

So could Dr. X explain his nervousness? "Oh, that's a good one," he said. During his first year he'd had trouble sleeping. "I got involved fairly early on with counterbiological warfare. At the time it wasn't really talked about. But within the context of Jason, not only was it talked about, but you really see the goods. And you see the goods not only historically of what happened in our country, but you see the goods on what's going on elsewhere in the world. And it's not abstract now, it's real." Knowledge of the goods has drawbacks. "One is that at night you can't get out of your head that these things are out there." Another is that he has children: "I'm politically liberal, because you somehow want to believe that everybody's nice. But then you see the proof of how evil certain things are. And then you realize that you've got detailed knowledge of that. And that's the part that gets a little scary. I know things that bad people would want to know. And I don't want anyone to come looking for it from me. The other part of that is—"

He stopped, then said, "You said scientists love to talk, and I guess that's true, here I am just jabber jabber," and changed the subject. He had told me earlier he was "proud, happy, and enriched by being a Jason." But he added, "I am somewhat freaked out."

Jason is usually capitalized, *JASON*, as though it were an acronym. I have no idea why—maybe because written like a name, it might be taken to be a personal name. I shall not capitalize it, partly because these pages are already full of acronyms and partly because writing it like a name implies the reality. That is, Jasons talk as though Jason is a separate person. Their pronouns of choice for Jason are not first person, *we*, but third person, *they* or *it*. They talk as though Jason were a sort of collective organism. But from the minute I first heard those ground rules, I understood I was going to have to get a sense of the collective without ever seeing it function. Even without the ground rules, the classification of their meetings meant I couldn't watch them work together. I would have to reconstruct the collective by talking to individuals: to show Jason through Jasons.

Nor could I read their institutional papers. Jason doesn't seem to have kept records in any systematic way, and those it does have are unavailable for various reasons, including classification and those ground rules. Except for Aaserud's archived interviews and his subsequent scholarly article, several meticulous government histories, and the few times that Jason found itself in the public record, what I could find of Jason's history depended on whatever records someone happened to keep and was willing to share; and after that I had only people's memories.

So this book is less a respectable history than a series of stories in more or less chronological order. The stories are chosen to answer those two questions—who Jasons are and how they've managed to last—plus the question I began with, why they're in Jason. The answer to the latter is: for lots of reasons, including curiosity and self-governance, money, secrets, and elite status. A fundamental reason is that the Jasons like each other. William Happer is a physicist at Princeton nearing retirement; he's been in Jason for twenty-eight years. Jasons stick because, he said, "I think they like to see each other." Academics usually keep to their own kind, so Happer the physicist says he especially likes working with the new nonphysicists like Dr. X—only he says Dr. X's real name—"who's just extremely open and cheerful and engaging. Where else would I get to know Dr. X?"

The most pervasive reason Jasons become Jasons is the reason the group formed in the first place: they feel responsible for how the country uses the products of science to defend itself. "We are scientists second and human beings first," Dyson wrote in *Disturbing the Universe*. A number of Jasons said the same, in nearly the same words. "We become politically involved," Dyson wrote, "because knowledge implies responsibility."

In the end, this book is a profile of the inheritors of the Manhattan Project, these scientists with their faith in the clarity—or at least the precise uncertainty—of pure science, their feelings of responsibility for its occasionally lethal consequences, and their willingness to navigate the accompanying political realities and moral messes. A career in science—with its competitive colleagues, its open disagreements, and the persistent unknowability of the physical world—is superb training in facing the possibility of being wrong. So along the way, the book also argues that the more convinced the government is of the rightness of its political decisions, the more it needs to hear the advice of its scientists. Jason does matter.

The Jasons

CHAPTER ONE

The Bombs

"You might decide not to do something; but the people anywhere else are just as smart as you are, and they're going to get there too."
—Sidney Drell, interview, June 2002

The idea that curiosity leads to disaster has an ancient pedigree. Pandora opened the gods' box and let loose all the evils of the world; the descendants of Noah built the Tower of Babel to reach heaven, but God scattered them and confounded their language; Icarus flew so close to the sun that his homemade wings melted, and he fell into the sea and drowned; Eve ate the apple of knowledge and was exiled from the Garden of Eden; Faust traded his soul for sorcery and spent eternity in hell. Saint Augustine, along with most medieval Christian theologians, considered *curiositas* a vice. The idea survived into the twentieth century about halfway intact. We believe that curiosity is the beginning of knowledge and especially of science, but we know that the application of science has led to disaster.

The so-called chemists' war was World War I. Its chief chemist was Fritz Haber, a German Nobel Prize–winner who figured out how to fix nitrogen—that is, pull it out of the air and use it to make both fertilizer and explosives—then directed the first full-scale chemical war, flooding the French trenches with chlorine gas. By the time the war ended, the British, French, and German military were using not only chlorine but also phosgene, hydrogen cyanide, cyanogen chloride, and bischloroethyl sulfide or mustard gas, among others; and chemical warfare had killed or injured a million soldiers.

World War II, by contrast, was the physicists' war. In the late 1930s and early 1940s British and American physicists developed radar: they'd send out a quick pulse of radio waves, which would bounce off a target and return; by timing the pulse's return, they could calculate the target's distance. For the first time they could locate an invisible, inaudible enemy; and airplanes, missiles, boats, and surfacing submarines became detectable from far away and in the night. Around the same time European physicists discovered that the nucleus of an atom could be split and in that split, or fission, would release extreme amounts of energy. Word travels fast and internationally among physicists, so almost immediately they all knew that the energy from fission—in particular, the fission of a certain form of uranium atom—would make extraordinary bombs. Some of the best of these physicists were in Nazi Germany and working on fission for the government; others were in Allied countries; the Allied physicists passed along news of the German experiments to the American government.

By mid-1941, Nazi Germany had taken over Austria, Czechoslovakia, Poland, Denmark, Norway, France, Belgium, Luxembourg, and the Netherlands, and it was attacking England. At the end of 1941 Germany declared war on the United States. And so in 1942, at the direction of President Franklin Roosevelt, a consortium of Allied scientists, eventually called the Manhattan Project, collected in groups all over America.

At the Metallurgical Laboratory at the University of Chicago, under the directorship of Enrico Fermi, they learned to build the reactors in which fission could be managed. In Oak Ridge, Tennessee; in Hanford, Washington; at the Radiation Laboratory at the

University of California in Berkeley; and at Columbia University in New York City, they learned to work with the two elements whose atoms were most congenial to fission. That is, they learned to separate the forms of uranium and to manufacture plutonium. And at Los Alamos, New Mexico, under J. Robert Oppenheimer, they learned to make the bombs. "Even before the end of our first year at Los Alamos, the emotional strain was apparent, the feeling that you've got to make that bomb, you've got to get it done; others are working on it; Germans are working on it; hurry! Hurry! Hurry!" wrote the wife of a Manhattan Project physicist. "We had parties, yes, once in a while, and I've never drunk so much as there at the few parties, because you had to let off steam you had to let off this feeling eating your soul, oh God are we doing right?"

In 1943 Herbert York was a graduate student in physics who was recruited to the Berkeley Radiation Laboratory, where he was trying to separate the fissionable form of uranium from the non-fissionable form. The secrecy at the labs was so great that uranium wasn't called uranium but "tuballoy" or "tube alloy": York wrote later that "saying uranium out loud had become equivalent to cursing one's mother." William Nierenberg was a graduate student at Columbia University, working on the same thing but with a different method.

Marvin Goldberger, known as Murph, was an undergraduate at the Carnegie Institute of Technology and a member of the enlisted reserve. He was called up and in 1944 was assigned to the army's Special Engineering Detachment (SED) of soldiers who happened to have scientific talent. "I was sent to Chicago," he said. "I worked on the design of nuclear reactors. We worked six days a week, sometimes seven. The only people we could talk to were the people that we worked with. You know, you couldn't go to the corner bar and tell what you were doing. I always said that I would have taken up with Mildred anyway, but it was unavoidable." Mildred Ginsberg, later Mildred Goldberger, was an assistant research scientist on the Manhattan Project, also at Chicago, and therefore was one of the few people whom Murph could talk to.

John Wheeler, a physicist at Princeton, was at Chicago working on reactors, too; later he went to Hanford and worked on manufacturing plutonium. Edward Teller, a physicist at the University of Chicago, was a consultant at Los Alamos; he calculated the effects of atomic bombs and thought that an explosion might ignite the atmosphere and incinerate the planet. Hans Bethe, a physicist at Cornell, was head of Los Alamos's Theoretical Division; he found Teller's calculations had some wrong assumptions, redid the calculations, and said the atmosphere and planet were safe.

Kenneth Case was an undergraduate at Harvard, and being a physics major, he said, he was sent to Los Alamos: "There were about eight of us who regularly ate, not in the mess halls with the soldiers, but at the lodge. And among those eight was Klaus Fuchs." Klaus Fuchs was then in the middle of his seven-year career of sending atomic secrets to the Russians. "So I must have had roughly three meals a day with him for a year and a half or so. I mean, he wasn't the funniest guy or anything. But I wouldn't have suspected." Case worked in the Theoretical Division calculating yields—that is, given different configurations of bombs, how much energy would be released. "I made the last calculation of the yield to be expected when the first bomb was dropped. And my guesstimate was eighteen kilotons, and the measurement was somewhere between fifteen and twenty, which was exact as far as I was concerned." A kiloton is the amount of energy released by a thousand tons of TNT.

Val Fitch was drafted out of Chadron State College in Nebraska and in 1944, like Goldberger, was assigned to the SED, but he was sent to Los Alamos, which had the majority of SED workers, or SEDs. The SEDs, though in the army, were assigned to work with the civilian scientists; they worked nonstop hours and made lousy soldiers. "The SED boys were terrible," wrote the wife of one Los Alamos scientist. "They couldn't keep in step. Their lines were crooked. They didn't stand properly. They waved at friends and grinned. Once, when a sergeant became irritated by his yawning, half-hearted crew and shouted, 'If you guys think I like this job, you got another think coming,' one of the SED boys offered to lead the drill in his place. He shouted orders in imitation of the sergeant's voice: 'Thumbs up, thumbs down. Thumbs wiggle-waggle.' Even the sergeant broke down and dismissed them." A civilian scientist intervened with the military, and eventually the SEDs were excused from reveille, calisthenics, and latrine cleaning; afterward, Fitch wrote, Saturday morning inspections "became devoid of spit and polish."

Fitch helped measure whether all thirty-two of the bomb's detonators would fire simultaneously. He designed the circuit that sent out the final pulse that would automatically trigger the bomb's first test. He climbed the hundred-foot tower on which the bomb was to be tested and helped hook cables to the bomb. On July 16, 1945, Fitch went to the main control bunker, six miles from the tower, stood outside, and watched the test.

The test, called Trinity, was at Alamogordo, New Mexico, and was of the plutonium bomb. The uranium bomb was too simple to need testing. The flash of light from the explosion reached the main control bunker in thirty millionths of a second, the blast wave in thirty seconds, Fitch said, and "I got up from the ground and watched the now famous mushroom cloud rise in the morning sky." Robert Oppenheimer watched Trinity from the same bunker as Fitch. Afterward he said he was reminded of the *Bhagavad-Gita*, of a line that has since become synonymous with the bomb: "Now I am become Death, the destroyer of worlds."

Isidor Isaac Rabi, a Columbia University physicist who directed the development of radar at MIT's Radiation Laboratory and consulted for the bomb builders at Los Alamos, watched Trinity from the base camp, five miles farther from the explosion than Oppenheimer and Fitch. "Suddenly, there was an enormous flash of light, the brightest light I have ever seen or that I think anyone has ever seen," Rabi wrote. "It blasted; it pounced; it bored its way right through you. It was a vision which was seen with more than the eye. It was seen to last forever. You would wish it to stop; altogether it lasted about two seconds. Finally it was over, diminishing, and we looked toward the place where the bomb had been; there was an enormous ball of fire which grew and grew and it rolled as it grew; it went up into the air, in yellow flashes and into scarlet and green. It looked menacing. It seemed to come toward one. A new thing had just been born." Rabi went to the trunk of his car, got out a bottle of whiskey, and passed it around. Fitch drove from the bunker to join Rabi at the base camp and had, he said, "a good shot from the bottle."

Enrico Fermi was at base camp with Rabi and hardly watched Trinity at all. He did a physicistly thing. He saw the light, waited for the shock wave, and then held his hand six feet from the ground and dropped small pieces of paper. The shock blew the pieces across the ground. Fermi measured how far they'd gone—two and a half meters—and from that calculated that the bomb's yield had been equivalent to ten kilotons. It was about twice that but, as Case had said about his own calculations, close enough.

Luis Alvarez was a physicist at the Berkeley Radiation Laboratory who had worked on radar—he invented the microwave early warning system and an airplane blind-landing system. Now, at Los Alamos, he worked on the bomb's detonators. He saw the Trinity test from a B-29 at twenty-four thousand feet. Another physicist at the Radiation lab, Wolfgang Panofsky, called "Pief," worked with Alvarez on devices to measure the bombs' yields. Panofsky watched Trinity from the B-29 with Alvarez. They were trying to measure the explosion's shock waves but were prevented by bad weather. "All we were able to do," Panofsky said, "was to make sketches of the mushroom cloud."

Three weeks after the Trinity test, on August 6, 1945, the United States dropped a uranium bomb on Hiroshima, and on August 9 a plutonium bomb on Nagasaki. Herb York said that the spectacle of Trinity was trivialized by the images of Hiroshima: "I'm getting preachy, but the really big effect was the pictures of Hiroshima and knowing what had happened. Not this big physics spectacle—that is impressive, but it's secondary to the knowledge of the relation between nuclear weapons and people."

The knowledge of the relations between nuclear weapons and people, Oppenheimer said, was a knowledge of sin. The Manhattan Project physicists, he said, "felt a peculiarly intimate responsibility for suggesting, for supporting, and in the end, in large measure, for achieving the realization of atomic weapons. Nor can we forget that these weapons, as they were in fact used, dramatized so mercilessly the inhumanity and evil of modern war. In some sort of crude sense which no vulgarity, no humor, no overstatement can quite extinguish, the physicists have known sin; and this is a knowledge which they cannot lose."

This sentence is still famous, I think, not only for its exquisite control of rhythm but also because it's an answer to what the rest of us want to know: what effect did that bomb have on its creators? Did physicists know sin? If sinning is doing something they knew would cause ordinary people terrible harm, then yes, they did. The bomb destroyed two cities and eventually killed 200,000 people in Hiroshima and 140,000 in Nagasaki, over half the population of both cities; the Manhattan Project physicists knew it would. The problem with assigning sin is this: what ended as terrible harm began as research whose motive was curiosity and whose outcome was unforeseeable.

Less than ten years before, two German chemists, Otto Hahn and Fritz Strassman, wanted to study how the elementary particles arrayed themselves into atoms of different elements. So they fired particles called neutrons at the nucleus of a uranium atom, expecting the nucleus to absorb the neutrons and turn into an atom of a heavier element. Instead, the element it turned into was lighter. The German chemists wrote to an ex-colleague—an Austrian Jewish physicist, exiled to Sweden, named Lise Meitner—for advice. In the last days of 1938 Meitner and her nephew, Otto Frisch, also a physicist, went for a walk in the snow to talk over the uranium nucleus, sat down on a log to do some calculations, and realized the nucleus had not absorbed the neutrons but had instead split in two. When it split, it lost some mass. "Now whenever mass disappears," Frisch wrote later, "energy is created, according to $E = mc^2$." M, the mass of the nucleus, was small; but c^2, the speed of light squared, was enormous, and so therefore was E, the energy. The chemists and physicists both published immediately; Frisch thought a good name for the process would be *fission*.

Physicists everywhere immediately understood that a fissioning uranium nucleus would release more neutrons and that these neutrons would, in turn, cause neighboring nuclei to fission, releasing even more neutrons and causing a chain reaction. Once that aunt and nephew sat on that log, calculating how an atom gave up its energy, an atomic bomb was inevitable and even obvious. As Richard Rhodes wrote in his definitive history of the atomic bomb, "The bomb was latent in nature as a genome is latent in flesh." Or as Val Fitch said, "Whether one likes it or not, these things are going to be discovered sooner or later, going to be used sooner or later."

So in their curiosity about the nature of the world, physicists had discovered that an atomic nucleus can be split and become an unparalleled weapon of mass destruction. Did they feel they'd sinned? Oppenheimer told the Los Alamos scientists, "The reason that we did this job is because it was an organic necessity. If you are a scientist you cannot stop such a thing. If you are a scientist you believe that it is good to find out how the world works; that it is good to find out what the realities are; that it is good to turn over to mankind at large the greatest possible power to control the world and to deal with it according to its lights and its values." Building the bomb had been an organic necessity, but by building it, physicists had known sin: in the curiosity-sin quandary, Oppenheimer seemed to come down firmly on both sides.

In diaries, reminiscences, interviews, and histories, the Manhattan Project physicists have said they didn't necessarily feel they'd known sin. But they also didn't necessarily feel good about their bomb, and they became preoccupied by how to contain it. They knew that the problem they now faced was, in their words, "getting the genie back in the bottle." The historian Daniel Kevles wrote in *The Physicists*, "The mushroom clouds had hardly cleared over Hiroshima and Nagasaki before atomic scientists started publicly, passionately, and persistently to plead for the international control of atomic energy." The effect of the bomb on its creators, if it wasn't sin, was an unrelenting sense of public responsibility. The issue was less moral than it was pragmatic; don't brood about it, just get it corked.

The Manhattan Project physicists had been trying to find ways to cork the genie before it ever left the bottle. They had argued back and forth among themselves and with the government about whether it should be used. Would the alternatives of fire-bombing and invasion kill even more ordinary people, Japanese and American? Would a staged demonstration of the bomb force the Japanese to surrender? Would the Japanese simply fight to the last man and never give up?

In June 1945, a month before Trinity, a group of Manhattan Project physicists gathered at the lab at the University of Chicago and wrote the so-called Franck report, arguing that the bomb should never be used but only demonstrated "before the eyes of representatives of all the United Nations, on the desert or a barren island." Those physicists lost the argument, probably for good reason. The report was classified secret and so was not read widely, but those who did read it considered it impractical and futile. The Franck report went on to predict that since German and Russian physicists already had "the basic scientific facts of nuclear power," and since uranium and plutonium were available to both countries, then nothing would prevent an arms race, "which will be on in earnest not later than the morning after our first demonstration of the existence of nuclear weapons."

Another attempt came the day after Trinity: around eighty-eight scientists at the Manhattan Project labs at the University of Chicago and Oak Ridge signed a petition to the president asking that the bomb not be used unless Japan refused an offer to surrender. The president apparently never saw the petition, and apparently neither did anyone else who could act on it. A month later, in September 1945, after the bomb was dropped on Hiroshima, the Atomic Scientists of Chicago formed to "support the immediate establishment of international controls over atomic power developments," the idea being that the arms race predicted by the Franck report might be headed off if the countries about to build the bomb all agreed to ban it instead.

Yet another attempt came the next month, after President Harry Truman introduced to Congress a bill, called the May-Johnson bill, that would create an Atomic Energy Commission to control nuclear materials and research. The new commission would be dominated by what was then called the War Department. Within two weeks the physicists went into overdrive, lobbying Congress and calling newspapers, protesting that the May-Johnson bill would, as Mildred Goldberger said, turn over "the whole kit and kaboodle to the military" and would therefore throttle both basic research and international cooperation. Mildred Goldberger was one of the scientists who signed the Chicago petition; the Goldbergers' friends were the lobbyists. "What they did," she said, "was to go to Washington—they had no money at all—they slept on sleeping bags in people's living rooms, and they taught physics, not to the Congress but to the staff members. And they won. They won." They didn't actually win for about a year, when the May-Johnson bill was allowed to die and the McMahon bill passed, putting the Atomic Energy Commission into civilian hands, thereby giving nuclear researchers a venue for international conversation.

And again: four months after Hiroshima, in December 1945, Manhattan Project alumni formed the Federation of Atomic Scientists to urge international control of nuclear weapons and the freedom of basic research. A year later it changed its name to Federation of American Scientists and then evolved into a multipurpose organization with several thousand members that analyzes, advocates, publicizes, and testifies on just about anything in national policy that's related to science.

The physicists could not control their genie. But for the first time in this country's history they became a national presence. Physicists' advice was sought on government committees and panels; they were asked to consult for industry; and their research was funded handsomely. They were heroes. In the same speech in which Oppenheimer said physicists had known sin, he also said, "Today, barely two years after the end of hostilities, physics is booming." John Wheeler told an interviewer, "I think the effect of the war was to make people appreciate, who were in the field of physics, that physics is important, that you can get money to do physics, you could do it in a big way, and you don't have to be a worm." Charles Townes worked on radar during the war and was now on the faculty at Columbia. "Suddenly physics had an aura," he wrote in his autobiography, "and physicists were popular at dinner parties."

Meanwhile, the Manhattan Project physicists went home to their universities. The younger ones continued their unfinished educations: for a career in physics, especially in academic physics, the price of admission is a doctoral degree—a Ph.D., the Philosophiae Doctor, awarded after doing research never done before—and usually a postdoctoral fellowship after that. The older Manhattan Project physicists went back to teaching and research. "We all felt that, like the soldiers, we had done our duty," Bethe wrote in an essay, "and that we deserved to return to our chosen life's work, the pursuit of pure science and teaching." Bethe went back to his teaching at Cornell. Graduate programs in physics, he said, were now overflowing with students.

"I said, 'the war was over,' and two weeks later I was at Harvard," said Ken Case. "I was interested in physics, not just in making bombs. No, but I mean it was boring after a while. Because we had all the techniques finally worked out as to how to calculate the yields. So we were just applying the same formulas in the end." After Case finished his Ph.D., he did postdoctoral research at Oppenheimer's new home, the Institute for Advanced Study; Case's adviser had neglected to write him a recommendation, said Case, but "Robert knew me from the war."

Val Fitch still had to finish college and headed off to McGill University, then later went to Columbia for graduate work. Herb York went to Berkeley to get a Ph.D. and had classes with Oppenheimer before Oppenheimer left for the Institute for Advanced

Study. Nierenberg went back to Columbia to get a Ph.D. with Rabi. Edward Teller and Enrico Fermi were faculty at the University of Chicago, where Murph and Mildred Goldberger were now in graduate school. Teller's students were theorists and Fermi's, experimentalists, and because Murph was a theorist, he should have been Teller's student. But, he said, "Fermi said to Teller one day, 'You know, during the war I've sort of fallen behind in theoretical physics. Maybe I ought to take a couple of students and try to catch up that way.'" Teller told a close friend of Murph's that Fermi was taking theorists as students. The close friend "went running down to Fermi's office and asked Fermi if he would take him as a student." Fermi would, so the close friend came back and told Murph, who "ran down the hall and asked Fermi, and he said yes. So we were Fermi's first two theoretical students."

One of Hans Bethe's new graduate students at Cornell was Freeman Dyson. Dyson, who was English, had spent the last two years of the war as an analyst for the British Bomber Command. He had known before going to Cornell that Bethe had been at Los Alamos, he said, but "I had not known beforehand that I would find a large fraction of the entire Los Alamos gang, with the exception of Oppy, reassembled at Cornell." Dyson felt lucky to be there. "There was endless talk about the Los Alamos days," he wrote. "Through all the talk shone a glow of pride and nostalgia. For every one of these people, the Los Alamos days had been a great experience, a time of hard work and comradeship and deep happiness."

If the Manhattan Project physicists, looking back, agree on anything, it's on Dyson's "comradeship and deep happiness." The Manhattan Project was a group of brilliant people who were greatly impressed by the brilliance of their colleagues. They were of all ages (though mostly young), classes, nationalities, cultures, and levels of fame. They were collaborative and egalitarian. They worked intensely and for an urgent purpose, and the work fascinated them.

They met people they were proud and pleased to have met. Fitch, who had come to Los Alamos from a ranch in the sand hills of Nebraska, found the European émigré physicists like Hans Bethe "a revelation."

The Manhattan Project experience affected even those physicists who hadn't worked on it. Edward Frieman got out of high school and enlisted in the navy, which then had sent him to college at Columbia: "But [even] as a young sailor, I knew something was going on at Columbia because we would see these odd characters floating around. But none of us knew really anything about it—it was all very secret stuff. When the United States dropped the atomic bomb, this was a tremendous eye-opener that such a thing existed, that you could do it. And so I somehow felt, not that I wanted to make bombs necessarily, but that the world had changed. And somehow I wanted to be a part of it."

The Manhattan Project disbanded, but work on bombs continued. Several years earlier, back at Los Alamos, Fermi had proposed a different bomb based not on fission but on its opposite, fusion. That is, instead of the nucleus of a heavy atom fissioning apart into lighter nuclei, the nuclei of light atoms would be pushed together until they fused into a heavier nucleus. Fusion is better than fission at freeing the energy in a nucleus; the fusion of hydrogen nuclei into helium in the sun makes sunshine. At the time that Fermi made his suggestion, however, the Manhattan Project was too busy with the fission bomb to worry about a fusion bomb.

But the Franck report's prediction had been dead-on, and by the end of the war the arms race had begun, driven by new political enmities. The Communist Soviet Union under Joseph Stalin was annexing the countries of Eastern Europe. In 1946 Winston Churchill gave his famous speech at a small college in Missouri: "From Stettin in the Baltic to Trieste in the Adriatic an iron curtain has descended across the Continent," he said, and behind that curtain the ancient capitals and states were all subject to an "increasing measure of control from Moscow." The United States announced that it would help any country trying to resist Communism, and the Cold War was declared. In 1949 the Soviet Union tested its own fission bomb.

Teller, a Hungarian refugee with a well-earned horror of the Soviets, wanted the United States to respond by building the fusion bomb—also called the hydrogen bomb, or just the Super—and pushed hard for it; Luis Alvarez agreed with Teller. Other Manhattan Project scientists—notably Oppenheimer, Fermi, and Rabi, who were advising the Atomic Energy Commission—wrote a report recommending the hydrogen bomb not be built: "The reason for developing such super bombs would be to have the capacity to devastate a vast area with a single bomb. Its use would involve a decision to slaughter a vast number of civilians." They called it "a weapon of genocide." Fermi and Rabi added, "It is necessarily an evil thing considered in any light." Both sides lobbied Washington strenuously. Washington noted not only that Stalin had the fission bomb but also that his best physicists, helped by Klaus Fuchs's information, were now working on the hydrogen bomb. In January 1950 President Truman ordered the hydrogen bomb built. Work proceeded at Los Alamos and at the Livermore branch of the Berkeley Radiation Laboratory.

Nevertheless, most physicists didn't think that building the hydrogen bomb was urgent. They didn't know whether it was even possible, and few liked it anyway. "Los Alamos had a difficult time recruiting people to work on the hydrogen bomb," Murph Goldberger said. "You know, they were interested and concerned but not that concerned. It wasn't World War III yet." John Wheeler, now back at his faculty job at Princeton, thought more senior scientists would sign up to work on the hydrogen bomb if they could

work, not at the weapons labs, but in academia. Wheeler himself
was in Teller's camp: "I am sometimes asked to name the most important peacetime use of nuclear energy," he wrote. "My answer is
simple: a nuclear device to keep the peace."

Wheeler set up a sort of satellite hydrogen bomb program at
Princeton, calling it Project Matterhorn. But he misjudged its attraction to senior scientists, and Project Matterhorn ended up being run with graduate students and new Ph.D.'s. One of these
Ph.D.'s was Ed Frieman, who was still in the navy and assigned to
Project Matterhorn but not yet cleared to know classified information. So Wheeler told Frieman, "We have a project here which I'd
like you to work on, but I can't tell you what it is. I want you to
think about two different elements, and I'm going to give them the
names gravium and levium. And I want you to figure out what the
surface tension is between gravium and levium." Frieman thought
to himself, "Oh God, I don't know what the hell I'm doing." When
he eventually got cleared, he found out he was working on the hydrogen bomb.

Both Richard Garwin (a student of Fermi's, then a junior colleague of Teller's, at Chicago) and Marshall Rosenbluth (a student
of Teller's) worked summers at Los Alamos on the hydrogen
bomb. Teller's design for the hydrogen bomb was complicated
enough that no one knew whether it could be made to work.
Rosenbluth worked on the theory, Garwin on the experiments.
Garwin set up an experiment that was not based on the design
"that Teller wanted," he said, but that worked as Garwin had set it
up to work. In October 1952 the hydrogen bomb was tested at a
Pacific atoll called Eniwetok. Nearly a year later, in August 1953,
the Soviet Union tested its hydrogen bomb. In March 1954 the
United States tested another one at another atoll, Bikini. In May
1957 Great Britain tested its own bomb. These tests were called
"shots," and ours were named: the test at Eniwetok was called the

Mike shot; the one at Bikini, the Bravo shot. Over the next ten years we did a few hundred shots.

Ed Frieman saw three of the shots, he said, "which, I think, to this day, still give me nightmares. The one underwater test, where the lagoon was full of Japanese ships that had been captured and others that the navy no longer wanted, and seeing something that appeared to me the size of the Empire State Building, standing on end with tons of roiling mud." Herb York was at Livermore and saw some of the other shots: "They were shooting bombs in Nevada. And there were the reports in the papers, people in Los Angeles had seen the eastern sky light up. So they asked me to go up to the Berkeley hills and see. There, three or four hundred miles away, big piece of sky lights up, just like dawn. It was a clear day and the Sierra Nevadas were between me and the site. And the whole spine was in silhouette against this bright—" He didn't finish the sentence.

In the spring of 1954, about the time of the Bravo shot at Bikini, the Atomic Energy Commission held hearings to decide whether Oppenheimer should keep his clearances. Some of his friends were members of the Communist Party; and he'd been making enemies in McCarthy-era Washington, partly by arguing against the hydrogen bomb. Teller testified against Oppenheimer, and even though Fermi, Bethe, and Rabi testified on his behalf, Oppenheimer lost his clearances and so could no longer advise the government. Physicists have disliked Teller ever since. Later, after Project Matterhorn ended, Teller asked Frieman to work for him at Livermore and Frieman refused: "I didn't want to work with Edward." And was that because of Oppenheimer? "Yeah."

But physicists' dislike of Teller had a second aspect, unrelated to Oppenheimer. Teller contradicted other physicists' pragmatic approach to the curiosity-sin problem: build the bombs only because the other side is going to build them anyway, then work to get them banned. Teller, to the day of his death, remained a strong and loud advocate of bigger, better, and more numerous bombs; he opposed the test ban treaties. He was credited with convincing President Ronald Reagan to launch the Strategic Defense Initiative, which physicists, impugning its basis in reality, called Star Wars. Teller, physicists thought, could ignore any physics that contradicted his own political beliefs; in fact, they thought he used physics to further his beliefs; they thought he was the original Dr. Strangelove. Because Teller's reputation was so bad for so long not only among physicists but also among the general public, and because American citizens are aware that theirs was the first country to have built the bomb and the only one to have used it, I wonder whether Edward Teller, whatever else he was, might also have been our collective scapegoat.

The Franck report had a preamble saying that after scientists had developed the atomic bomb, they could no longer "disclaim direct responsibility for the use to which mankind had put their disinterested discoveries." The scientists' sense of responsibility never left them. Murph Goldberger, Freeman Dyson, Ken Case, Val Fitch, Luis Alvarez, Pief Panofsky, Bill Nierenberg, Ed Frieman, Marshall Rosenbluth, Charles Townes, and Richard Garwin all became Jasons. Teller, Wheeler, and Bethe were three of Jason's four senior advisers; Herb York was their first sponsor.

Richard Garwin, on first impression anyway, was easy to underestimate. He looked like an unremarkable middle manager with a receding hairline and glasses. He talked in simple, declarative sentences and withheld the ordinary emotional, human connections. His own description of himself was disconcertingly rational: "I was good at finding out how to do things that wanted to be done. So I went to Los Alamos. I learned everything there was to know about nuclear weapons in a week or two, and I realized that if they were going to make a hydrogen bomb there was this experiment that needed to be done. So I started building that experiment. And then I realized they were going to have a whole lot of bombs. These are very nasty things—you don't want to be blown up by nuclear weapons. So I developed an interest."

Garwin went on to have a long career in industry and academia; his reputation among intelligent physicists has been for his intelligence. But if he's famous for one thing, it's for advising every administration since Eisenhower's, regardless of party and whether they want him to or not, about the technical problems of national security. In the spring of 2000 he gave a speech to graduating physicists at Berkeley: "While you are enjoying your work, occasionally seized by the divine passion for solving a problem," he told them, "put some time aside regularly to think about your role in preserving your society against ignorance and unreason."

CHAPTER TWO

Jason Is Born

"Modern military technology is like a bear. You can grab a bear and run as fast as it does, and it won't turn around. But let go, and it will bite you. The world is more complicated than we like to think."

—John Archibald Wheeler, interview, 1991

Herb York got his doctorate in 1949, did physics research for a few years, and then in 1952 became the first director of the new national weapons laboratory, the Livermore branch of the Berkeley Radiation Laboratory that was renamed, years later, the Lawrence Livermore National Laboratory. At Livermore, York was in charge of designing and building new hydrogen bombs that had, he wrote, "the smallest diameter, the lightest weight, the least investment in rare materials, or the highest yield-to-weight ratio."

Meanwhile the countries that had those hydrogen bombs—the United States, Great Britain, and the Soviet Union—settled into the Cold War's terrifying gamesmanship. Each side's move depended on what it thought the other side might be doing. All sides were testing lighter and bigger bombs with yields no longer in kilotons but in megatons, and developing the missiles to shoot the bombs from one continent to the other, halfway around the world. All sides were, at the same time, alternately approaching and avoiding some sort of agreement about controlling the bombs or at least stopping the bomb tests. All sides offered bases for agreement and then refused them, made concessions and withdrew them, planned conferences and canceled them, and, all the while, negotiated the politics of national pride and special interests back home. The situation was a mess.

In late 1957 the mess suddenly got worse. On October 5 the Soviet Union sent into orbit a satellite called Sputnik that weighed 184 pounds. It beeped, sending out little radio signals, so nobody could miss it. "Everyone who was at all knowledgeable," York said, knew that "the rocket needed to launch a satellite and the one needed to throw a warhead of the same size at a target sixty-five hundred miles away are virtually identical." The bigger the rocket, the bigger the warhead it could carry and the farther it could go.

President Eisenhower tried to reassure the country. Then a month later Sputnik was joined by Sputnik II, weighing 1,120 pounds, a few pounds of which was a dog. Eisenhower went on television to say we weren't in a race to get to space and two weeks later had a mild stroke. A few days after that, with lots of publicity, the United States launched the biggest rocket that it could then launch, the Vanguard, designed to carry a satellite that weighed 3.5 pounds. Vanguard lifted off a few feet, fell back, and blew up, all on national television. Not until the end of January 1958 did the United States finally succeed in answering Sputnik with a Jupiter rocket and a satellite that weighed 31 pounds. The weight of hydrogen bomb warheads at the time varied, but they were closer to a thousand pounds than to 31.

This disparity in size had roots in World War II. Both American and Soviet rockets were the evolutionary progeny of the German V-2 rockets: after the war, both countries divvied up the German V-2 scientists, technicians, and techniques. The Soviets saw satellites as military and so pushed their evolution toward heavy satellites and correspondingly large rockets. The United States decided to split the satellite program into military and scientific components and so pushed its evolution in two directions. The scientific component wanted light earth-observing satellites, which could be carried into orbit by smaller rockets. The military component was diffused throughout the armed services, but one branch was working on heavy spy and early-warning satellites, along with the requisitely large rockets. When Sputnik went up, the military's rocket-satellite system was actually well under way, but because the system was so highly classified, no one could say so.

National uproar ensued: commentators declared our education in science and math hopelessly behind, the phrase "missile gap" appeared regularly in news stories, our cities were threatened with annihilation, we would have to build fallout shelters, and we'd better watch the skies. Jack Ruina was an engineer at the University of Illinois: "While we were developing a five-pound little satellite that was going to do some small things, the Russians put up a hundred-pound—I forgot the number—satellite that the Americans could see. I remember living in Illinois, and there it was, every night. The next satellite had a dog. And there was a joke, the next one's going to be a cow, and then they're going to put up several cows, and the joke was—I don't know if you heard this—it'll be the first herd shot round the world."

Both Congress and the press demanded noisily and repeatedly that the country immediately build bigger rockets. The people who built bigger rockets reacted predictably. "Arms builders everywhere," York said, "who had any plausible connection whatsoever to rockets and missiles and satellites were coming to Washington with all kinds of ideas, all of which were intended to save the republic from imminent disaster." Congress and the press also demanded more defense research and better scientific education. Somebody had to sort through all the demands and possibilities. So once again the government found a use for scientists and quickly set up a flock of institutions for getting their advice.

A month and a half after Sputnik, on November 21, 1957, President Eisenhower revived the "sleepy" (Garwin's word), "moribund" (Goldberger's word) President's Science Advisory Committee, or PSAC (pronounced, York writes, "pea-sack"). PSAC was made up of sixteen scientists who were neutral, independent, and employed elsewhere, who had no particular constituencies or financial interests, and who reported to the president. PSAC busied itself with questions, among others, of antisubmarine warfare, chemicals in food, and arms control. Its chair was Isidor Rabi, and its members included Hans Bethe. It had unusual power and influence. In December 1957 Herb York left Livermore for Washington and joined PSAC. "I was whatever I was, thirty-six," York said. "I got to know the president and secretary of defense and all those people, because they were honestly looking for advice and help."

Three months after Sputnik, in January 1958, the secretary of defense set up a new science-based agency, reporting directly to him, named the Advanced Research Projects Agency and always called ARPA. ARPA was a remarkably small, remarkably open-minded home for high-risk, high-payoff (ARPA's words) military projects. The secretary had in mind "anti-missile missiles and outerspace projects," but wouldn't rule out "highly speculative types" of weapons. In March 1958 York decided to accept the job of ARPA's chief scientist. His decision to join the Defense Department, wrote Charles Townes, "surprised a lot of his colleagues. I remember asking Herb at about that time why he did it, and he joked, 'Well, they paid me a whale of a lot of money!' It was a way of saying that whatever the reasons for working in Washington, enjoying science or enhancing a professional reputation were not among them."

Fifteen months after Sputnik, in December 1958, the secretary of defense set up another science office to oversee the design, development, testing, and evaluation of all of defense research, including ARPA. The office was called the Office of the Director of Defense Research and Engineering (ODDR&E), and York, after only a few months at ARPA and in spite of his youth, was asked to be its first director: "So I dealt directly with the White House, directly instead of through the secretary of defense. It just was the special nature of the time." As DDR&E, York had not only access to power but power itself. "I now had full authority over appropriations and expenditures," he wrote. " 'The power of the purse' is just about all the power there is."

And so for the next decade, as never before or since, money was spent. The charted line of the federal money spent on research and development between 1940 and 1960 looks like a mountain you could climb. For the line from 1960 to 1968, you'd need ropes and harnesses; it's nearly vertical. The money going into research doubled; the money going into basic research tripled. The number of doctorates earned in physics doubled; the number of universities offering doctorates in physics also doubled. (An irrelevant, but significant, aside: of 223 universities total between 1962 and 1964, the same 21 received over half the federal funds and awarded half the doctorates. The funding system was meritocratic, not egalitarian.)

What came out of this bounty was, said Jack Ruina, "a little wild." Ruina, by 1960, had left the University of Illinois to become one of York's assistant directors at ODDR&E; the next year Ruina left ODDR&E but stayed in the Defense Department and became head of ARPA. "Left to its own devices," he said, "the military, for a short period, was a little bit nutty. Russians making hundred-megaton bombs? We're gonna make thousand-megaton bombs. One general was saying we have to have a base on the moon because the military should be on high ground. And industrial people—I'm not going to say it was deliberate—they went right along with it. Somebody asked me, don't I have any regrets? I had a budget of three hundred million dollars, which was like one and a half billion in current dollars. I had no oversight—Congress never questioned about giving us what we wanted. I look back, why was I so timid? Why didn't I do really bold things? Because I didn't realize these were special times."

"It's hard to reconstruct now the sense of doom," said Wheeler, "when we were on the ground and Sputnik was up in the sky." Wheeler was worried that if the Soviets got the edge militarily, they would become "adventurous," he wrote, and trigger another world war. He understood immediately that the government needed advice on technical areas like missiles, "where the outlook of a physicist could be useful."

At the time Wheeler was at Princeton, working on the structure of spacetime on the atomic scale—he called it "quantum foam"—and fusion for the hydrogen bomb. The split between nearly theological physics and hard-eyed defense is classic Wheeler. On the one hand, he coauthored the standard book on gravitation/general relativity; he named "black holes"; he worked on the anthropic principle, which explores the possibility that the universe evolved so that we could observe it. On the other hand, Wheeler—along with Edward Teller and Luis Alvarez—was one of a small but public and influential group of physicists supporting the development of the hydrogen bomb; his colleagues' disapproval, he said, was "deeply troubling." "Basically John Wheeler believed in nuclear weapons," said Ed Frieman. "I mean, every once in a while he would say things that would absolutely scare me to death." Back during the Manhattan Project, while Wheeler was at the Hanford laboratory trying to manufacture plutonium, he had learned his younger brother was missing in action on the Italian front. "It was a year and a half before his body was found," said Wheeler. "Well, I can't call it his body—his skull and his skeleton. I always think to myself that if I'd only gotten going sooner on making plutonium for a bomb, I could have saved his life."

When Sputnik flew, Wheeler talked to two of his Princeton colleagues: Eugene Wigner, a Manhattan Project physicist, and Oskar Morgenstern, an economist with a specialty in game theory, a way of describing and predicting human behavior with mathematics. "And we thought the thing to do," Wheeler said, "was to get a way for new people to get involved in defense problems."

The "old people" involved in defense problems—the scientists on PSAC and PSAC's many subject-specific panels—were pretty much the same physicists who had built the bombs and advised on their control. They were a tight group: they'd gone to the same few graduate schools and were faculty at the same few universities; a large fraction of PSAC came from the same side of the Charles River in Cambridge, Massachusetts—that is, from Harvard and MIT. "Actually, we all know each other," said one PSAC member, an MIT physicist named Jerrold Zacharias, who had worked on both radar and the Manhattan Project. "People always think that because the United States has a population of one hundred seventy million and there are a lot of people in the Pentagon, it all has to be very impersonal. Science isn't. It's just us boys." Princeton—the home of Wheeler, Wigner, and Morgenstern—also had a tradition of government service, but MIT was "enormously more dependent on defense money for its budget than Princeton was," Wheeler said. "You would find a larger percentage of people here at Princeton feeling defense issues are important and it's an honorable thing to get involved in them. You'd find a larger percentage here than you would along the River Charles." In any case, by "new people," Wheeler meant specifically not the old River Charles crowd.

The way to get new people involved in defense problems, Wheeler thought, was to set up an enterprising ARPA-like entity, not as a smallish agency, but as an interconnected set—"a whole campus," he said—of national laboratories. A few months after Sputnik, Wheeler proposed what he called the National Advanced Research Projects Laboratory: "And we talked up the idea and wrote a proposal—and to whom did it go?"

It went to just about everybody. Over the next few months Wheeler, Wigner, and Morgenstern made successive and changing proposals, wrote letters, made calls, held meetings, and leaked to the press. A *Life* magazine photograph shows Wheeler, Wigner, and Morgenstern, along with a young Murph Goldberger wearing startling socks, all sitting around a room in leather armchairs. They renamed the idea the National Security Research Laboratory, then the National Security Research Initiation Laboratory. The idea evolved into a single lab full of scientists acquainted with defense problems who would "initiate" ideas. The ideas that "turned out to be good," said York, who was then still at ARPA and one of the people getting their letters, "would go to industry" for development. By this time, the president's science adviser, James Killian of MIT, was involved and opposed; he wanted to give PSAC and ARPA a chance to get on their feet before starting a new laboratory. Someone proposed, as a dry run for a research initiation laboratory, a summer study.

Summer studies had been invented by the River Charles crowd. Academics have summers free, and some part of the government would fund a group of them to study a particular problem— antisubmarine warfare, for example—for a summer and write up a report. Sometimes a summer study would expand beyond the summer; different summer studies were done by different groups that, when a study was over, dispersed. Richard Garwin said, "There was nothing that couldn't be done, according to the MIT crowd, by a summer study." The results of summer studies were reputedly uneven, ranging from remarkable to mundane. Zacharias, who supposedly began the studies, said modestly, "Summer studies and some are not."

The Wheeler-Wigner-Morgenstern summer study of a research initiation lab cost $50,000 and was funded by ARPA. "And I bought the summer study," said York. "I mean, I was the one official who actually both thought it was a good idea and had money." To keep the study's character publicly opaque, Wheeler named it Project 137—1/137 is physicists' so-called fine structure constant, a number that quantifies the strength of the electromagnetic force on the atomic scale. A physics story goes: a physicist gets to heaven and God asks him if there's anything he'd like to know, and the physicist says, "Yes, please. Why is the fine structure constant 1/137?" God gives him the explanation, and the physicist says, "No, that's wrong."

Project 137 met in the summer of 1958, from July 14 through August 2. "We got the green light for twenty-two scientists for two weeks," said Wheeler. No one seems to know who chose the scientists. Wheeler said the criteria were experience in defense problems, practicality, ingenuity, and motivation. "It's like any club," Wheeler said. "You see somebody—'why, he would be wonderful to add'—you talk it up with other people, and if the idea gains enough currency, then a formal nomination gets made." Murph presumed the choice was made by "the big three—Wigner, Wheeler, and Morgenstern—with a certain amount of knowledge of people who were becoming involved in the consulting business." York said they just picked the best scientists: "I mean, we didn't count down from the top and get the best fifty, but they were right up there at the top. I don't know who took the main initiative. I was involved in it closely. But I didn't pick them."

The scientists invited to Project 137 were younger than Wheeler, and much younger than the River Charles crowd, but still not all that young; they were, on average, in their early forties. Murph Goldberger was thirty-five; he'd gone to Berkeley for a postdoctoral fellowship and was now a professor at Princeton. Val Fitch, thirty-four, and Sam Treiman, thirty-three, were also at Princeton; Kenneth Watson, thirty-six, was a professor at Berkeley. Of the twenty-two Project 137 scientists, two were from industry and four were at the national laboratories: an unconventional friend of York's, named Nicholas Christofilos, was at the Livermore lab. Sixteen of the twenty-two were academics—three were chemists, two were mathematicians, one was an economist, the rest were physicists; six were from Princeton. Not one was from Harvard or MIT. All likely had top secret clearances.

Project 137 met at the National War College in Washington, D.C. "It was a quiet spot with a nice library," Wheeler said. "The army, navy, and air force told us in top secret about their hottest problems." The twenty-two scientists, he said, "came up with twenty-two ideas per problem." Most of Project 137's two weeks was spent listening to the Defense Department's briefings. The briefings were intense, Murph said, "and they went sort of from morning to night. And we went out to dinner and then to bed. And it was hotter than hell in Washington in the summer, as it always is. Some briefings were dreadfully dull and some were quite interesting."

The briefings that most impressed the participants included those on the difficulties of detecting enemy submarines and the need for communicating with ours; the need for the army to be able to fight wars with both conventional and nuclear weapons;

and the increasing superiority of the Soviet Union in missiles and in radar and electronic equipment. Wheeler remembered briefings on guerrilla warfare: "trick ways to detect enemy movements," he said, "chemical sensors based on the ability of an insect to detect a smell." Murph remembered "one involving biological warfare, which involved infecting rats with terrible disease and driving them against an enemy. It made everyone sort of sick to the stomach." Watson thought the briefings "were intended to be rather shocking. Most of the attendees, like myself, were not very familiar with much of this."

Watson said that the primary purpose of those briefings was education. Nevertheless, Murph said they'd listen to briefings, then talk over coffee, "and physicists have a tendency to try to solve problems." Fitch said, "As you know, when someone is talking, you're not spending much of the time thinking about what they're saying. You're thinking about the problems they might have presented to you. So there's a foreground subject but in the background your brain is turning over. And then we had time later to do back-of-the-envelope calculations to see if we really had any ideas or not."

Whether or not they were supposed to solve problems, they did. The resulting report, *Identification of Certain Current Defense Problems and Possible Means of Solution,* included short appendices, each devoted to a single problem and anywhere from one to several pages long, that were written by individual or small groups of scientists—whoever, said Wheeler, "got really stewed up about something." One of these appendices outlined a plan for converting civilian radios to battlefield radiation detectors. Another was pages of equations on the feasibility of a nuclear-powered, high-speed torpedo that would roam the oceans, preventing enemy submarines from shooting missiles at U.S. cities. Wheeler didn't think he wrote the report's main body, but the introduction and summary read as though he did: "Many members of Project 137 were deeply disturbed and others even shocked by the gravity of the problems with which they found themselves confronted. These men in number constituting less than 1/4000 of America's scientific community have been stimulated to intense activity by what they heard. . . . The group senses the rapidly increasing danger into which we are inexorably heading."

Under all the briefings and calculations was the question of the future of the National Security Research Initiation Laboratory. Project 137 was considered a success, so the idea of a research initiation lab seemed to have had wide support from everyone—participants, hosts, and sponsors. The exact form this lab would take was still a little vague. The Project 137 report outlined two more possibilities with two more names, neither of which sounded much like Wheeler's original possibilities. Wheeler presented the secretary of defense with Project 137's recommendations, including the ones for the lab, and eventually, he said, "that proposal became so much alive that I was asked to be head of it. They wanted it to be at Princeton. So this was not an empty proposal, it was a very concrete thing." And though Wheeler didn't normally gesture, here his hands framed a building—not just a piece of advice, but a real thing, a tangible thing. An aside: Wheeler's drawing for this building is in the shape of a pentagon. "There was this building," he said, "would I run the show in it?"

No, he would not run the show. York asked first Wheeler, then Murph to set up and direct the laboratory. Murph said, "Wheeler refused to become the director. I refused to become the director. The notion of attracting people out of academia to work full time in something like that was not wildly attractive." Neither Murph nor Wheeler wanted to leave physics. Murph said Eugene Wigner told him, "You know, if you take this job, people will forget what you've done in physics maybe fifty years sooner." Wheeler said,

"That was a most agonizing time for me, because I wanted to—
'how the world works' is the poor man's way to state what I was
always interested in. But here I'd been pushing for the project. I
wanted to have an automobile collide with me and break a leg, so I
would have a clean way out. Made me feel like a worm to say no,
but I did."

If Wheeler wouldn't stay with the cause, neither would Wigner
and Morgenstern. York said he wasn't sure an initiation lab was going to work anyway: "Although I talked in a welcoming way with
Johnny about those things," he said, "I was doubtful we could
make that work. Much too expensive in terms of what we wanted.
And a little bit precious. Johnny's ideas about what they could contribute were probably, well, beyond reality. Having been the director for five or six years of an applied physics laboratory, I didn't
really believe that a laboratory the size of Livermore could possibly
work in a free-wheeling fashion."

In other words, as York knew, national laboratories have buildings and staffs and scientists to support, programs and missions to
protect, and political winds to weather. The scientists a lab hires
have a vested interest in keeping the laboratory going, so any advice that contradicts the lab's programs, missions, and political
winds will undermine the lab's future, not to mention the scientist's job. Nor can a national lab's scientists necessarily choose what
research interests them or, once they have chosen, easily change interests. None of this conduced to "free-wheelingness," and the idea
was a nonstarter. By early spring of 1959 Project 137 was over.

By now York had changed jobs from ARPA to DDR&E, with
its "power of the purse." "See, what I think did come of the
Wheeler-Morgenstern-Wigner approach," he said, "was the notion
of free-wheeling people who were very good and who worked in
an intimate connection. When we finally got the picture of Jason
as a summer consulting group, I'm quite sure that in my mind, I

said, 'That's the way to do it. Not the Wheeler way. This is the way to do it.'"

A government wanting scientists' advice on, say, the feasibility of a particular system of missile defense could ask scientists who are nearest—that is, who are in the defense industry or on the Defense Department's various advisory committees. But those scientists, like the national lab scientists, would have something political or financial to gain or lose and might hedge their advice accordingly. Disinterested advice comes best from independent scientists, like those on PSAC, outside the government and outside industry— that is, from scientists employed in academia whose livelihoods will not depend on the advice they give.

The government had a number of ways to get advice from academic scientists. The River Charles approach was a group that met for a summer study on a specific subject at a specific site for a couple weeks, then dissolved. The ARPA approach was to fund individual academics (among others) for multiyear research at their own universities. The PSAC approach was to put a team of (mostly) academics on call year-round without necessarily keeping them on site. The Project 137 approach was to take a team out of the academy and install them in a national laboratory. The national labs' and defense industry's approach was to pay individual academics to consult during the summer. In what York called "a process of serendipity and successive approximations," two subgroups of people combined these approaches and, said York, "cooked up the notion of what we now call Jason."

The first subgroup was made up of three academics, and their approach was to go into the summer-consulting advice business on their own. The three were Murph Goldberger, Kenneth Watson, and Keith Brueckner. They had met in the late 1940s, doing graduate (Brueckner) and postdoctoral (Watson and Goldberger) work at the Berkeley Radiation lab. By the late 1950s, about the time Sputnik went up and Wheeler was thinking up research initiation labs, Goldberger, Watson, and Brueckner had taken faculty jobs. During the school year they worked on physics, generally on trying to make sense of the flood of new subnuclear particles coming out of the new atom-smashers called cyclotrons. During the summers, first at Los Alamos and later for the flowering, post-Sputnik defense industry—particularly for Convair—they consulted.

Summer consulting had become common, especially among physicists, after the Manhattan Project. Unlike summer studies, consulting is done individually and usually for a national laboratory or for industry. Universities don't mind their faculty members consulting—ties with industry and national labs help graduates get jobs and make bequests more likely—and often give them a day a week to do so. Scientists do summer consulting because universities generally pay salaries for the nine months of the academic year only. In 1960 the average public school teacher made $5,174 and congressmen just under $25,000; in 1964 full professors of physics typically earned between $12,000 and $15,000. In 1957, said Watson, "I went to Berkeley as a full professor and my salary was between ten and eleven thousand dollars. I was a little bit appalled—I went from Indiana to Wisconsin and I took a cut in salary, went from Wisconsin to Berkeley and I took a cut in salary—and Edward Teller said, 'Look, your salary's irrelevant, it's your consulting that will give you your income.'" Some physicists could, by doing summer consulting, at least double their salaries.

At Convair, Goldberger, Watson, and Brueckner were working on a government contract to inquire into the effects of nuclear explosions in the upper atmosphere. "And we were a little irritated by the fact that we were doing all the work," said Murph, and Convair was passing on to the government "nothing but what we had done, and was getting paid twice as much." The three decided to deal directly with the government and cut out the middleman. "Keith Brueckner and Ken Watson and I began talking about forming a company that would contract directly with the Department of Defense," Murph said, "and not screw around with intermediaries."

One motive for forming a company was, of course, financial. Another was to be in control of what they'd work on. "When you consult for a company, you do what they want you to do," said Watson. "And it was generally stimulating, but it wasn't what we were in control of." A third motive, Wheeler's motive, was to get new faces into the cadre of Manhattan Project science advisers. Murph and his colleagues were beginning to serve on national science advisory groups—defense panels, PSAC panels, PSAC itself. "One of the things that got us moving was," Murph said, "as we gradually began to be invited on to various committees, we always saw these World War II warriors and thought maybe it was time for the youngsters to step up to the plate and share some of the responsibility."

They decided to call their company Theoretical Physics, Incorporated—"a lousy name," said Watson—and hired an attorney to draw up articles of incorporation. "All physicists all knew each other so we were in an excellent position to get a supergood staff," said Watson. "Essentially our idea was to get a lot of smart people together but to do it in a private company." Lest anyone be disturbed by physicists operating on the profit motive, let me point out that Theoretical Physics, Incorporated never happened anyway.

Goldberger, Watson, and Brueckner dropped the idea of Theoretical Physics, Incorporated after talking to the second subgroup, Marvin Stern and Charles Townes. During the war Townes had worked on radar, then had gone to Columbia. Meanwhile he had also been on a number of government and military panels and concluded from these experiences that "Washington badly needed scientists." Moreover, at Columbia Townes had met the Manhattan

Project science advisers, like Rabi. "I observed what these people
had done," he said, "and thought, 'What a fine thing it is.'" So when
Townes was asked to become the vice president in charge of research for the Institute for Defense Analyses, he was inclined to accept.

The Institute for Defense Analyses is generally called IDA, except by John Wheeler who calls it by its full name and pronounces
the plural in "Analyses." IDA, which also hosted Project 137, is
what was then called a federal contract research center, or an
FCRC, one of a number of private, nonprofit companies formed
after the war whose staffs were scientists and whose trustees were
mostly academics. FCRCs were another nonacademic way for government to get science advice, and in fact, FCRCs' sole customer
was the government—in IDA's case, the Joint Chiefs of Staff and
the secretary of defense. Townes agreed to work at IDA for two
years, he said: "I thought two years was the minimum time to do it
well and I could stand that much." He said he thought of it as his
public duty.

Townes said he learned about Theoretical Physics, Incorporated
in a conversation with Marvin Stern. Stern, though he agreed about
the conversation, said he'd never heard of the company. Stern is a
little hard to pin down. He was a mathematician who worked for
Convair, he'd worked for Herb York either at ARPA or at ODDR&E,
and he seemed to specialize in knowing physicists: he had invited
Teller, Wheeler, Wigner, and Morgenstern to consult at Convair
and, he said, "I already knew Goldberger, Watson and Brueckner."
Stern had received one of Wheeler's early proposals for a research
initiation lab and had attended Project 137. After Project 137
folded, Stern joined a committee to advise IDA's directors. Like
everyone else, Stern noticed that the Manhattan Project advisers
were aging and recommended to IDA that it contract with a group
of young academic scientists to help solve defense problems. "I was

talking about particularly uninitiated guys who had never heard about a problem in defense," Stern said. "My motives were twofold: take the bright young guy, expose him, he may get some ideas for himself and his lab or for the government; but my number-two motive was to suck him in, get him interested in problems in defense, like the dog that discovers the virgin woods, 'Trees, trees, they're all mine!' You know what I mean?"

Whether the conversation between Townes and Stern was about Theoretical Physics, Incorporated or about Stern's advice to IDA about trees isn't clear; but the outcome was that Townes got the idea of an independent, free-standing, ongoing group of smart young physicists contracting through IDA to spend summers working together on defense problems. "When I interacted with Townes, he understood," said Stern. "He had great—I'll use the word—cultural perception of the potential significance of something like this." Townes thought Theoretical Physics, Incorporated should be modified in one particular way: it should be nonprofit. A nonprofit doesn't need to shape its advice according to what will turn a profit and so can be more objective. "I felt that was important," Townes said. "The idea was not to make money but to do something useful."

So Townes talked to his superiors at IDA and to the secretary of defense, and they too thought it was a good idea. Then Townes and Stern went to Los Alamos to meet Theoretical Physics, Incorporated. Townes told Goldberger, Watson, and Brueckner that if they contracted through IDA, they would be "much more on the inside and trusted to give objective advice, you see, rather than selling something to the government." Watson said that Townes called on their patriotism. Murph said he "leaped at the notion that we could have sort of a parent organization that would take care of us." Brueckner noted that "of course becoming a division of IDA gave us an immediate source of financial support." Watson added,

"And also excellent contacts in the government." Townes said, "Why, they were very pleased with the idea. I didn't have to persuade them. I just had to explain what the possibilities were, and they accepted it very quickly."

IDA was actually only a conduit through which support could be channeled. In pursuit of someone with money to channel, Townes talked to Herb York, who had known Goldberger, Watson, and Brueckner from Berkeley and who was now DDR&E. York liked the idea and gave ARPA approval to fund the group; the contract between ARPA and IDA was for $250,000. In addition to setting up a support channel, Townes also had to ensure the group's autonomy. He negotiated with the Defense Department over whether the group would—as do most science advisory bodies—work on a single project assigned to it by a specific entity; or whether it could work on projects of its own choosing in varying areas and for varying agencies. The Pentagon wanted the group badly, but "they wanted them under the Pentagon wing," Townes said. "I had to struggle with that to see that they were more independent."

ARPA called the new group Project Sunrise. "The Pentagon had a machine that generated names of projects and operations," said Murph. "You would describe a project and push a button, and they would cough up a series of names." Nobody liked the sunrise. "It sounded a little bit presumptuous," Watson said.

"It had the wrong feeling," Brueckner said. "We were not creating a sunrise."

"We weren't going to develop a new superbomb or something," said Watson.

Murph went home from Washington and told Mildred the name, he said, "and she said whatever the moral equivalent at that time was of 'that sucks.' She said, 'You should call yourself Jason.'"

Mildred had seen IDA's annual report, which was decorated by mythological figures and a colophon that looked like a Greek temple. She was admiring this annual report, she said, "when Murph came back from D.C. and told me they'd be named Sunrise, a name with no character at all. He said, 'So what's a better name?' I picked the name Jason primarily because it was a recognizable name, not just a collection of letters. It's Jason and the Argonauts looking for the Golden Fleece. It was the group and the quest. Plus it's a nice word, easy to say."

Plus it's a little off the wall, a group named Jason. It begs for explanation. Besides July-August-September-October-November, Herb York wrote in the first draft of his autobiography that Jason was named "apparently by Mildred Goldberger after their family dog." The Goldbergers told York they never had a dog named Jason—"we had a dog named Jack," said Murph, "but not Jason"—and York corrected the final draft. The authors of the official and thorough history of ARPA say in a footnote that Jason was motivated by patriotism and money both, inspiring the authors to remark, "This no doubt contributed to the designation accorded Jason by the non-scientists in ARPA, namely, 'the golden fleece.' "

Did Mildred know the whole Jason myth? Did she know that Jason survived his adventures and got the Golden Fleece only because his wicked lover, Medea, a sorceress, rescued him every time he got into trouble and then put the dragon guarding the Golden Fleece to sleep, so all Jason had to do was pick up the fleece? "I knew that," said Mildred. "I knew Medea did the work. I didn't do it on purpose, there was no malice aforethought. Jason would be thought of as young heroes. Jason wouldn't be a bunch of middle-aged men smoking cigarettes and drinking coffee. He'd be tall, well-muscled, blond. I was just thinking about a group setting out on a voyage with a destination and winning."

Did Murph's cofounders see a connection between their group

and the Jason myth? "You're looking for too much," said Watson. "It's just a name."

"All I know is Murph told me it came from Mildred," said Townes, "and I said, 'Well, that sounds fine to me. Does everybody accept it?' And they said, 'Sure, that's fine.'"

On that Golden Fleece charge: it doesn't stand up, though getting numbers to back that is difficult. Going with Murph as the source on Jason in 1960: at a per diem fee of $50 for maybe thirty days in the summer and maybe six days a month for maybe ten months, Jasons could add to their university salaries somewhere around $4,500. It wasn't small change, but it wasn't golden.

Once he'd gotten Jason set up, Townes also had to get Jasons cleared. "Some of them were young and had spoken out in radical-sounding ways," he said, "and so it wasn't easy to get them the kind of clearance we felt important." The difficulties were less with the Jasons than with the times. "We'd been through the McCarthy era," Townes explained, "and very suspicious attitudes had been taken— 'these guys are kind of Communist and we shouldn't have them inside on anything.' It's a very natural attitude. One can understand it, even if it isn't right." At any rate, Townes told the clearance-issuers that the Jasons were loyal Americans, and he said, "they may have studied it in whatever secret ways they do. But then they agreed." Townes said the clearances were unusually broad and inclusive.

Then Townes was politically clever: he also proposed that the new group have a committee of senior advisers, three-fourths of whom were clearly conservative. "I wanted to be sure [the group] had the respect of the White House and the Pentagon," he said. "And these senior people were well known in the Pentagon and the White House. So John Wheeler we got. And Eugene Wigner was not only an outstanding scientist, he had come from Hungary and the Communist world he knew very well and he was dead against them. And Teller was clearly anti-Communist, an outstanding scientist, he was very, very well known in the government. And then Hans Bethe. And Hans Bethe was a little more liberal but very highly respected in the government and an outstanding scientist. I think, for the Pentagon and the White House, feeling that the senior advisers could straighten us out if we did something wrong was probably pretty important."

If anyone is to be considered the father of Jason, I'd vote for Townes. He talked Theoretical Physics, Incorporated out of the profit motive, talked a willing York into funding an independent group, talked IDA into backing it, talked the Defense Department out of controlling it and into giving it broad access, and made it politically palatable. Townes's understanding of what Jason could be and therefore how it must be set up was hard-headed and clear-sighted and in the end assured its independence.

He was proud of what he did for Jason. In the two years he spent at IDA, he said, "I would say that is undoubtedly the most important single thing that I did." And was Jason something new? "I guess it was," said Townes. "I guess it was. In terms of getting academic people advising the government, I think it was brand new."

The Glory Years

"All you have to do is have a concept and people will give you a hundred million dollars. The scientists who brought you the atom bomb, brought you radar . . . they can do anything. Just give 'em the money and priority."

—Jack Ruina, interview, 2002

Jason was born in the middle of the opulent, venturesome post-Sputnik period that the ARPA historians called "the golden era" and Jack Ruina called "special times." Robert Oppenheimer, asked at his clearance hearings to explain his earlier arguments against the hydrogen bomb, said that part of his reason was that no one had debated the atomic bomb until after they'd made it: "When you see something that is technically sweet, you go ahead and do it and you argue about what to do about it only after you have had your technical success." Technical sweetness is an elegant match between form and function, between a technology and its purpose. It evokes a feeling intense enough to amount to a compulsion. As a motive, it is probably universal in scientists but is not at all limited to them. Technical sweetness seemed to have been the golden era's motto.

The young Jason pulled itself together fast. After agreeing with Townes and Stern, the members of the now-defunct Theoretical Physics, Incorporated had met to discuss Jason, "and in one day's discussion," Murph said, "we sort of fleshed out the concept and I walked out of the room to go to the john and when I got back it turned out that I was the chairman." Murph agreed to be chair, in spite of having turned down leadership of Project 137, because Jason was only part time: "I wasn't about to stop working in physics," he said. Watson, Brueckner, and Murph—along with a fourth, younger member of Theoretical Physics, Incorporated, Murray Gell-Mann—appointed themselves the first steering committee.

Next, Jason had to get the right people. With the cooperation of York and Townes, the new steering committee held a recruiting meeting on December 17, 1959, at IDA in Washington, D.C. They began with a list that Theoretical Physics, Incorporated had already drawn up, Watson said, of people who were "smart and interested in applied science." Recruiting them turned out to be easy; perhaps twenty-two invitees came, which Brueckner said was at least three-fourths of the people asked: "It was a very elite operation. It was an honor to be asked." The invitees were "young and full of beans and very patriotic," Murph said. They were also, as advertised, smart and creative: seven—roughly a third of them— later won Nobel Prizes. The normally unimpressed ARPA historians wrote that ARPA "would have been hard-pressed to recruit replacements of equal calibre."

Among those at this first meeting were Val Fitch, Hal Lewis, Sam Treiman, and Ed Frieman. They, and the rest of what Murph called the primeval Jasons, "were the friends of Watson, Brueckner, and me," he said, "and by some quaint coincidence they all turned out to be physicists." Most of the physicists were theorists, Murph said, but "a non-trivial number" were experimentalists. When he says they're all physicists, he means they're exclusively physicists—no chemists or biologists or "any of those outsiders," he said. In spite of their professional purity, they were chosen for their abilities as generalists who could move into any field and solve the problem from first principles. That first meeting gave them the chance to hear briefings, and decide whether or not to join. One of the briefers was George Kistiakowsky, President Eisenhower's science adviser.

Kistiakowsky was a card-carrying River Charles warhorse. He was a Russian who fought on the losing side in the Russian revolution, fled to Germany where he got a Ph.D. in chemistry, was hired at Harvard, then went to the Manhattan Project where he directed a crucial part of the development of the plutonium bomb; he was the civilian scientist at Los Alamos who talked the military into loosening up on the SEDs. After the war he went back to Harvard and research, began consulting for the government, and after Sputnik became Eisenhower's second science adviser. While he was science adviser, he kept a remarkably forthright diary: "Thursday, December 17, 1959 11:00 AM—12:00 M: Met at IDA headquarters with the 'bright young physicists,' a group [called Jason] assembled by Charlie Townes to do imaginative thinking about military problems. It is a tremendously bright squad of some 30 people to whom I talked off-the-cuff on the general problems facing us. . . . I emphasized that new ideas which didn't suffer from complexity and wouldn't involve billions for development were most important and then spoke of the general objectives as being the creation of a secure deterrent and of an effective small force for use in limited engagements. Although feeling slightly hangoverish from last night, I think I did a reasonably good job and my remarks were followed by a very active discussion. This was followed by a luncheon which was just like other luncheons."

On January 1, 1960, Jason became an official entity, its duties and limits spelled out in ARPA's project assignment to IDA for Project Sunrise. ARPA told Jason to hire young, smart scientists; to solve technical problems; to point out science that academics weren't developing but that the military might use; to analyze but not to experiment. It anticipated that "minimum expenditures will be made for computers [and] assistants." The prohibition on computers was a relic of the time, the beginning of the computer age when, said Hal Lewis, a primeval Jason, "we lost students to computers— they got mesmerized and forgot to do physics. You didn't want this to be turning into a computer buffs' organization." The rationale for the prohibition on assistants, Lewis said, "was that Jason should be the work of the absolutely top physicists and one didn't want to use their names when graduate students really did the work." ARPA ended its project assignment by saying, "the Secretary of Defense and the Director of Defense Research and Engineering have expressed their keen interest in the concept of the group and have specifically directed that the project be pursued vigorously."

Between January 1, 1960, and the first summer study that June, some Jasons dropped out and others joined. One early and unusually young Jason was Steven Weinberg, age twenty-seven: he had been Treiman's student at Princeton and was then at UC Berkeley. "I wasn't that much younger than Gell-Mann, but Gell-Mann was a prodigy," he said. "But compared to most of the people who had founded Jason like Goldberger or Watson or Brueckner, I was considerably younger. I regarded them as senior figures and myself as a very junior figure."

Another Jason was Sidney Drell, age thirty-four, called Sid, his name usually pronounced as though it were all one word, sidrell. A year after the bomb dropped on Hiroshima, Drell graduated from Princeton, where he'd done his senior thesis with John Wheeler. About the time of the first fusion bomb test, Drell had finished his doctoral work at the University of Illinois, where his advisers and professors had been Oppenheimer's students; and he had joined the faculty at MIT, where he'd met Hans Bethe. A year before Sputnik flew, Drell became a professor at Stanford, where he worked on the theory of quantum electrodynamics. Drell hadn't wanted to do anything except "try to be a good physicist," he said, but "I had the privilege of knowing very well some of these people, and I admired what they were doing." So when Townes called Drell one night in early 1960 and asked him to try Jason out, Drell agreed: "one's honored to be called by Charlie Townes." Besides, he said, "once you knew what an atom bomb was and then once the ten kilotons went up to megatons, any physicist thinking a little bit had a pretty good idea of what the devastating destructive potential was. I couldn't escape it. I couldn't isolate myself from the world, that's what it was."

Once Jason had the right members, it figured out how to run itself so that it wouldn't go out of business—that is, so it would remain useful to ARPA and still be independent and collegial enough to attract and keep its academic members. The lifestyle it set up never really changed. It had an administrator hired by IDA, a physicist named David Katcher. Governing, done by the chairman and steering committee, was straightforward and informal. "You can't be tight with a bunch of academics," said Treiman. "Nobody wanted a tight structure, and Goldberger has a loose style anyhow." Murph said that Jasons didn't need much governing. "I was the leader, but only almost in a nominal sense; there was certain scut work that I had to do," he said. "But when you choose extremely smart people and unleash them on things that they can take a real interest in, you don't have to do much leading. Jason was an enormously—I don't want to say democratic—but everybody was involved almost equally. There was just a high degree of consensus."

The senior advisers turned out to be mostly, in Murph's words, "window dressing"; Townes said they were "somewhat figureheads," though reassuring and encouraging. Wheeler and Wigner came to meetings occasionally; Jasons in the opposing political camp argued with them. Bethe doesn't seem to have come to many meetings, and no one remembers seeing Teller at all: "A lot of Jasons didn't have much admiration for Teller," said Brueckner. Watson added, "All Teller would do is develop bombs."

The steering committee had two big jobs, one of which was to select members. New members were invited from among their colleagues, collaborators, or students—that is, as Murph said, people whose characters and work were familiar. "It always happens with old boys' networks," Treiman said. "Your vision is a little narrow sometimes. You don't know the other good people so well, so you don't think of them. Those are all human traits." New members weren't official until voted on by all Jasons, and then they were given an informal trial period, which David Katcher said "was minimal, because it was such a goddamn nuisance to get cleared that you didn't go into it unless you were pretty sure you wanted it and they were pretty sure to have you."

The steering committee's other big job was to supervise the process by which Jason chooses its studies. Every spring, somewhere around the end of April, Jasons would meet for a weekend in Washington, D.C., to hear government sponsors talk about their general problems. The steering committee would then discuss—primarily among themselves, sometimes with the membership—which problems would be the best matches. Every summer, from mid-June to the end of July, Jasons would gather out of town for a six-week meeting, hear briefings detailing the chosen problems, and go to their rooms to work separately or together on answers. Every fall, around the end of November, they would meet again for general education in problems they might want to consider for the spring. Eventually they also met during the winter, for two weeks in January, to do small studies or preliminary work on summer studies.

Asked how Jason decided which studies to select, Katcher said, "All sorts of ways." One way was that Jasons who had been on PSAC or other science advisory panels, Katcher said, would know which technical problems were nationally critical and recommend those problems to the membership. Another way was that the sponsor would say what was important to them and Jason would agree to work on it. Or the sponsor might "just paint the whole picture," Katcher said, "and then through the questions Jason's possible contribution would develop." The painting-and-questions process continued through the summer, the study being defined as it went. The point was to find the best match between the sponsor's problems and the Jasons' abilities. In short, the process of selecting what to study seems to have been one of those messy, iterative, interactive, human processes that turn out to be the most efficient and effective way of doing business.

In the end, the studies selected were those that Jasons signed up to work on; Jasons voted with their feet. The unusually broad classification clearances that Townes had negotiated meant that all Jasons could hear all briefings and then, said Treiman, choose voluntarily from the menu: "You were never, never directed." Undirected choice from the menu was the only way academics would work, just as following their own intellectual leanings was the only way they could do their research.

The broad clearances also meant that all Jasons could work with all other Jasons and thereby benefit from everyone's expertise. In these early days "everybody got reports from everybody on what everybody was doing," Katcher said. Most of the Jason studies, said Katcher, were done collaboratively: "I think one of the pleasures of the summer study was that they worked together." Treiman said, "It's a chatty bunch."

From the start, Jason's studies followed ARPA's missions. ARPA's biggest mission was a program, called Defender, to develop a defense against ballistic missiles. Over the years the mission has changed names—Anti-Ballistic Missile defense (ABM), Strategic Defense Initiative (SDI or Star Wars), and National Missile Defense

(NMD)—but the question is always the same: how to figure out what's being shot at us and how to defend ourselves.

The Defender program, in line with ARPA's high-payoff, high-risk character, had a binary personality. On the one hand, the ARPA historians say that Defender was "credited with major contributions to ballistic missile defense": Defender developed the phased array radars that could track many missiles simultaneously and fast and that became the basis for all subsequent missile defense systems. It ran the tests that made the measurements that allowed incoming decoys to be discriminated from warheads. It developed short-range, fast interceptor missiles, so we could wait until the last minute before firing. On the other hand, the ARPA historians also say that Defender had "a slightly flaky, if not outright bizarre sort of image"—studies on magnetic barriers and antigravitation devices—though nothing in this category, Ruina said, was "taken seriously or very far." Defender was costing upward of $100 million, roughly half of ARPA's budget. Much of Jason's first summer was spent on Defender problems. If ARPA in particular and science advising in general were in the midst of their glory years, so was Jason. Jason's work on Defender, said the ARPA historians, quoting an ARPA director, "resulted in major contributions."

One of these contributions came immediately, during Jason's first summer meeting, held at the Berkeley Radiation lab which was now called the Lawrence Berkeley National Laboratory, from June 1 to August 15, 1960. Around twenty Jasons came. Sid Drell, who had been unsure whether he wanted to join Jason, worked with another Jason on a particular Defender problem and found himself drawn in; he called it "an example of entrapment."

The background of the Defender problem is simple: a missile launched by another country should be detectable by the heat—that is, by the infrared radiation—coming off its plume of flame. New post-Sputnik spy satellites had wide fields of view and infrared detectors, and if the satellites were high enough, they could see the missile as it rose above the atmosphere. "The missile plume is very hot and bright above the atmosphere for quite a distance," said Drell. "And the idea was, this would give you thirty-minute launch warnings because it takes thirty minutes for a missile to fly from where it's launched in the middle of the Soviet Union to the United States." Once you knew it was coming, you could fire your own missiles, said Drell, so "if the guy knew that you would know he's launching a missile and could fire yours, he might be deterred."

This is the basic outline of Cold War gamesmanship, or measure-countermeasure: one side launches a warhead-carrying missile; the other side puts up a satellite that detects missile plumes' infrared radiation; the first side might then hide the plumes' infrared with a nuclear explosion, and this was the problem posed to Jason. "Someone raised the problem," Drell said, "that as a precursor to an attack on us, the bad guys could detonate one or a few nuclear weapons high up in the atmosphere." The explosion would create a cloud of nitric oxide molecules, called NO, which radiate so much infrared that the plumes' infrared would be swamped. "So the question was," said Drell, "if you have a high-altitude explosion of a nuclear weapon and it makes a lot of NO, would that cause a big enough cloud to last long enough that we wouldn't see the missile attack launch and we wouldn't get the early warning?" Drell and the other Jasons calculated the amount of nitric oxide, the size of the cloud, the duration of the cloud, whether the wind would blow it around—"a terrific, interesting problem," Drell said. It turned out that an explosion large enough to swamp the plume had to be so many megatons as to be impractical.

The problem was not only interesting in itself, of course, but it also had implications in the real world of policy. The air force had a countermeasure, a satellite called MIDAS, designed to issue an early warning of an attack by detecting the infrared signatures of missiles. The Jason study, said Drell, allowed the air force to judge that MIDAS, though it turned out to have other, unsolved technical problems, was unlikely to be blinded by what was called "redout"—like a blackout, only infrared—and was therefore a feasible approach to early warning. For Drell, that combination of science and policy was the "entrapment." He felt his work was needed, he said, "and you go on from there and you get involved in other problems and pretty soon you're trapped."

Jason's second summer, in 1961, switched coasts, from Berkeley to Bowdoin College in Brunswick, Maine. The steering committee had decided to alternate coasts, east and west, in order to alternate geographical inconvenience. Bowdoin College was in driving distance from John Wheeler's summer house on the coast, on High Island. Jasons met on campus and lived in nearby rented farmhouses. "One Sunday we arranged a picnic right here in High Island, a clam bake," Wheeler said. "Actually of course it was not so much clam as lobster."

That summer Jason took up another of Defender's measurecountermeasures, this one to detect not a missile being launched but the warhead it carried re-entering the atmosphere. If one side could detect incoming warheads with radar, then perhaps the other side could confuse the radar by surrounding its warheads with decoys; if so, then the first side must learn to distinguish real warheads from decoys, best done as the warheads re-entered the atmosphere. To protect against the rigors of re-entry, warheads are surrounded with so-called re-entry vehicles. The size of the re-entry vehicle, Murph explained, is "some measure of how much nasty stuff it might have in it." One way to distinguish harmless decoys was by their size.

The Jasons knew of a difficult technique astronomers used to measure the size of stars, and they knew of a variant of that technique invented by a British radio astronomer and radar expert named Robert Hanbury Brown and his colleague, Richard Twiss. The Hanbury Brown-Twiss variant, like the astronomical technique, allowed sharp resolution of bright stars and therefore accurate measurement of things like re-entry vehicles. Murph, Watson, and Hal Lewis wrote a report called "HB-T Primer," explaining it all. The Hanbury Brown-Twiss technique had been discovered only a few years before the Bowdoin summer study and, said Lewis, the "government simply didn't believe us. So we thought we had to do an experiment."

They did the experiment the following year, in 1962, Jason's third summer, back on the other coast, in Berkeley. Though experiments had been discouraged in ARPA's original assignment, this one was cheap and easy, done "on the rolling hills of Stanford," Murph said, "with otherwise mild-mannered high energy physicists." Murph and Lewis, who were theorists, enlisted the help of three experimentalists: Val Fitch and two other early Jasons, Courtenay Wright and Leon Lederman. "It was Courtenay Wright, Leon Lederman, and myself who did the work," said Fitch. Murph, apparently, just watched. "Murph was not an experimentalist," said Fitch. "I don't mean that in any pejorative way."

Lewis said they got "an old radar set and sent some poor physicist out on a hillside nearby." Fitch said, "It was a very hot summer. We had radar receivers and we had a radar transmitter." Physicists on one hillside beamed radar at another physicist on another hillside, who held a reflector and was therefore a re-entry vehicle. "Val and Leon were running around with antennas in their arms and screaming back and forth to each other, telling each other what they should do," said Murph. The radar hit the reflector, bounced back, and was picked up by the receivers; and because the physicists knew the size of the re-entry reflector, they knew that the measurement they then calculated was accurate. "We had a great time," said Fitch, "and it worked. It was a new kind of physics for us."

The fourth summer, in 1963, was back on the Massachusetts seacoast, at Woods Hole. The National Academy of Sciences had a meeting house there, what the East Coast's old-money called a "cottage"—that is, a mansion with huge windows and a wrap-around porch—right on the water. By now Keith Brueckner had replaced Charles Townes as vice president for research at IDA and therefore had dropped off Jason's steering committee. Jack Ruina had become the head of ARPA.

That summer Jason was working on the high-risk side of Defender's personality, on what are now called directed energy weapons. *Directed energy* means beams either of light or of charged particles that are focused so tightly that none of their energy dissipates before the beam hits the target. Beams of light were, of course, lasers. Beams of charged particles were made of either electrons or protons and were essentially directed lightning bolts. The charged particle beam program was called Seesaw. "It was whether you can use a particle beam, earth-based, to form a beam through the atmosphere and destroy an incoming warhead," said Ruina. "It was very sophisticated physics involved. All kinds of questions about the stability of the beam, aiming the beam, how it spreads, whether it wiggles. Just like a fire hose. Jasons were really the first group to analyze it." Brueckner said that Seesaw became one of Jason's "hobby horses": "We worked on it over and over again." Everybody said that Seesaw had serious problems with the laws of physics. Ruina said that having Jason work on Seesaw meant that the "physics was going to be as good a physics as you can get. But I'll tell you, you didn't have to go that far to see that the whole weapon didn't make much sense."

"The particle beam was a creation of N. Christofilos," wrote the ARPA historians, "a brilliant and eccentric thinker who seemed to be able to mesmerize fellow physicists." Christofilos's charged particle beam, the historians said, had "a sort of 'Buck Rogers' death ray image," and Project Seesaw, despite not making sense, was "the most enduring specific project ever supported by the Agency." What kept it alive was that no one could prove that it wouldn't work, and every time another obstacle arose, they said, Christofilos came up with another solution, "each such solution requiring further intensive investigation." They quoted Herb York saying, " 'Nick was a remarkable idea man. The ideas were usually not good, but they were really remarkable in that they were the kind of ideas that nobody else had.' " The ARPA historians add in a footnote, "One of Christofilos' 'not good' ideas was to build a large aircraft runway across the entire U.S., coast to coast, so that the Soviets could never catch most of the SAC aircraft on the ground at the same time."

Nicholas Christofilos was one of the few national lab scientists to be a Jason. His ideas ranged from technically sweet to cosmically goofy. His photos from around 1960 show a dark man with slicked-back hair and a certain presence; he looked like a portly movie star. He looked approachable and compelling, as though he could talk you into things. "He was fat, sort of small and fat," said Freeman Dyson. "And two hundred percent Greek. And always talked with tremendous enthusiasm about whatever he was doing. I used to say that Christofilos was probably as productive as all the rest of Jason put together. As a result of him, Jasons made a rule that they don't allow you to work two days in twenty-four hours. I liked him very much." Jasons admired Christofilos—or at least enjoyed him—to the point of mythology. He must have returned the feeling: Murph said Christofilos had named his son Jason.

Nick Christofilos was born in America but raised in Greece, where he got an engineering degree, though not a Ph.D. He worked for an elevator maintainence business, which he later owned and which, when the Germans occupied Greece, became a truck repair business. The work wasn't that interesting so he read papers in physics journals, particularly papers on the design of accelerators, machines like the cyclotron in which atoms were collided so hard they splattered, and their debris analyzed for subatomic particles. He came up with a design for one of these accelerators, wrote it up in a letter, and sent it to the Berkeley lab. Unknown to Christofilos, his accelerator had already been invented, so Berkeley set the letter aside and forgot it. Two years later Christofilos wrote a second letter describing yet another, more complex accelerator; the Berkeley lab couldn't figure out what he was saying and set this one aside, too. Two more years went by, and Ernst Courant at Brookhaven National Laboratory published a paper inventing the accelerator that Christofilos had described in his second letter; Courant called it the cosmotron, and Brookhaven later built it. Christofilos happened upon Courant's paper and wrote a third letter that said that he'd already invented the cosmotron. The Berkeley scientists found his paper in their files, Courant wrote, but they "had examined it superficially and dismissed it as one of the many crackpot letters that laboratories get. They and we were most embarrassed, and we published a letter in the *Physical Review* acknowledging Christofilos's priority." Christofilos was paid for his trouble and was offered a job at Brookhaven that, in 1953, he took.

Almost immediately Herb York, then the head of Livermore, invited Christofilos to join him. "He was working on something of direct interest to us, and he was such an interesting person," York said. Livermore, like Los Alamos and Princeton, was trying to control fusion. Christofilos designed a fusion machine, called Astron, that would shoot a beam of electrons into hot hydrogen, heating it further until its atoms would fuse and release their energy. Astron's electron beam could also be used to study the behavior problems of the beam weapons of Project Seesaw. Like Seesaw, Astron never quite worked, but Christofilos loved and defended it, and besides no one could prove it wouldn't work someday.

Meanwhile Sputnik went up, and like Wheeler and Teller, Christofilos was worried sick about the Russians. "Nick came into my office basically frantic," said York. "And he was desperate about how to intercept intercontinental missiles." Christofilos's plan for intercepting missiles was "a huge cloud of energetic electrons held in place by the earth's magnetic field," said York, which would fry any missiles coming through it. To test his idea, Christofilos proposed exploding nuclear bombs in the atmosphere—this was before any test ban treaties—thereby creating the electrons, which would then be constrained by the magnetic field. Cristofilos's proposal in October 1957 roughly coincided with York's move from Livermore to ARPA. "ARPA is the only place that could pick up something like Christofilos's idea and support it," said York. "So ARPA order number one or number four—something like that, it's a very low number—is to do the Argus experiment."

The Argus experiment was actually three highly classified 1.2-kiloton explosions in the late summer of 1958; it worked, and it still sounds crazy. The electrons were trapped in the earth's magnetic field and traced its outline, creating auroras where the field dipped toward the earth. Whether the electron field would fry missiles seems unclear; it certainly fried satellites. Someone leaked it to *The New York Times*; the *Times* and then the rest of the media called it the greatest or largest or grandest scientific experiment ever done. "The men of Project Argus spun a veil of electrons around the earth, boldly using the atmosphere and nearby space as their laboratory," wrote *Time* magazine. "It almost seemed impudence."

Christofilos was invited to Project 137, the Jason precursor, and was one of the early Jasons. In Jason, Christofilos seems to have worked mostly alone. His solo projects had names like "Preliminary Thoughts on a Space Fleet" and were classified secret. "One of Jason's big jobs," one Jason said, "was trying to show that various crazy ideas of Nick Christofilos wouldn't work." Christofilos's most famous Jason project was a scheme to communicate with submarines. Like the Argus shots, the scheme was at the same time highly classified and, eventually, highly public; and as a result, much was written about it, and much of what was written was vague or contradictory or partisan and in any case generally untrustworthy. The Jason précis of the scheme tends to run along the lines of Jack Ruina's: "I forgot what it was called, his super-low frequency—he was going to set aside the whole state of Wisconsin and Minnesota and maybe half of Canada to be an antenna."

Some facts are ascertainable. The first name for the scheme was Bassoon. Christofilos first proposed it in 1958 during Project 137, just before the first Argus shots. The submarines were most likely the ones carrying nuclear missiles; communicating with them was difficult because the radio frequencies normally used to communicate over distances fade out, or attenuate, under water, and these particular submarines prefer the deep ocean. The problem could be solved if the submarines regularly came to the surface, but then they'd be obvious targets. Christofilos's idea was to use radio frequencies so low they weren't even VLF, very low frequencies, but ELF, extremely low frequencies. The lower the frequency, the less water attenuates it.

Christofilos's exact proposal is hard to come by: the Project 137 report was declassified after twelve years, but its eighteen pages describing Bassoon are still unavailable. Two years after Project 137, however, Christofilos wrote another report, declassified in 1972; by now, Bassoon was called Sanguine. The report proposed a frequency of 25 hertz: extremely low frequencies mean extremely long wavelengths, and 25 hertz corresponds to a wavelength of 7,400 miles. The longer the wavelength, the longer the antenna needed to transmit it; at a cell phone's frequency of a billion hertz or wavelength of ten inches, its antenna need be only a few inches long; at Sanguine's 7,400-mile wavelength, its antenna would be 8,500 miles long. The antenna was to be a loop, each end of which would be buried in the earth: a current traveling along the antenna to one end would run deep into the earth, then back up through the other end. The loop in turn would broadcast that signal with its thousands-of-miles-long wavelength. The ELF signal would bounce between and be guided by the conducting rock in the depths of the earth and the conducting part of the atmosphere called the ionosphere—the idea originally of another immoderate inventor, Nicola Tesla. The ELF signal goes right around the earth and hundreds of feet into the ocean. Besides nonattenuation, ELF signals have two other relevant characteristics: they have the virtue of being unlikely to be disrupted by Argus-like nuclear explosions in the atmosphere; and they carry little information. Christofilos said this system would transmit six words per minute. He figured it would cost $138 million. Such an uninformative one-way signal is essentially a beeper; Jasons call it a "bell-ringer." "The signals would be only of emergency-type signals, like 'Go to hell,'" said Ruina. Or, said a Jason, "'My God, we've an atomic attack. Go and mutually assure destruction.'"

Christofilos wrote around eight Jason reports on Bassoon/Sanguine. They were all classified confidential or secret. He seems to have worked alone on them; the reports list no other authors. The Jasons nevertheless all know about this project—it's one of the few they mention without being asked—though they are vague on the details, either because they don't know them or because they don't remember which ones are classified. Murph said that Bassoon/Sanguine was "one of the most important things that came out of Project 137" and the only thing with any "real application." So it was definitely built, right? "There were certainly wires laid," Murph said, "but I don't want to talk more about it." But it was built? "It was built," he said. Was it used? "I can't answer that," he said.

Christofilos was not a typical Jason. He was never an academic. His intuition about physics was good, but he was more of a cut-and-try inventor than an analytical theorist. A month after his last Jason report, "Interim Sanguine Systems," on September 24, 1972, at age fifty-six, Christofilos died. He's buried in a Livermore cemetery owned by the Independent Order of Odd Fellows. What Jasons seemed to like about him was his combination of unprepossessing education, astonishingly large ideas, and immovable loyalty to those ideas. Astron and Seesaw, though they never succeeded, didn't die until Christofilos did. And some of the ideas actually worked. Not only was a later version of Bassoon/Sanguine used, but the Christofilos/Courant cosmotron, also in a later version, was the accelerator Val Fitch and Leon Lederman used to find and characterize new particles and win Nobel Prizes in, respectively, 1980 and 1988.

By 1965 or 1966 the sheer mountain of government research funding had reached new heights, and ARPA's budget for Jason had doubled to over $500,000. The Jason experiment was clearly working well, even splendidly; Jasons were enjoying themselves and the government was finding them useful. The ARPA historians said Jason's early years included its most significant contributions. An unpublished in-house history said that for Jason's first five years, "by any realistic measure," its productivity was "exceptional."

Jason's usefulness to ARPA went beyond the Defender program to ARPA's secondary mission, the Vela program of detecting nuclear tests. In the early 1960s the best way to test whether a new bomb would work was to detonate it. The Soviet Union had been mostly testing huge nuclear bombs, and the United States had been testing bombs of just about every other size that were deliverable by just about any method, including by hand. Before the nuclear countries would sign a treaty to stop testing, they needed to be able to reliably detect anyone exploding bombs—either abrogating the treaty or cheating on it—in the sky, under water, or under ground. Jasons did studies on the effects of nuclear explosions in the upper atmosphere, like "Radiation Escape from a High Altitude Fireball," and under water, like "Water Waves from Large Nuclear Explosions."

By 1963 enough progress had been made in detecting explosions that the United States, the United Kingdom, and the Soviet Union were finally able to sign the Limited Test Ban Treaty, promising not to test nuclear weapons in the sky—either in the atmosphere or in space—or under water. The nuclear countries didn't promise not to test under ground, however, because the seismic waves from underground tests couldn't be discriminated from the waves from earthquakes; or if tests were done in a big enough cavern, the seismic waves might be muffled. Banning unverifiable tests seemed silly. So ARPA worked on seismic verification; Jack Ruina said ARPA was supporting "almost all of seismology in the United States." And Jasons did studies like "Seismic Signals from Nuclear Explosions."

The ARPA historians said that while Jason's work on Defender involved some of the most difficult technical problems in the Defense Department, its work on Vela provided the key to the government's decision to sign the Limited Test Ban Treaty. Ruina said that the Jasons themselves were more capable and more disinterested than most of the government's science advisers. "I think people looked at their material as being the most authoritative on the subject by far," he said. "Jason had more to offer in those areas than probably anybody." A subsequent ARPA director, Charles Herzfeld, said, "I have always considered them and still do consider them, the best, the most highly skilled, brightest scientific talent that the Defense Department has available."

The Jasons, who were unapologetically happy to belong to an elite group, cheerfully agreed. "There was tugging and hauling about the assessments of prospective members but not about our elitist principles," said Sam Treiman. "At the time, we were very cocky." Jasons felt they could walk into an unknown problem and contribute to its solution, said Murph, whose sense of humor had a particular slant, "and since it's a bunch of extremely smart guys, they ordinarily can. That's part of the attraction, to show that the thing some poor bastard labors over can be dispatched with a few clips and swaths of a sword by a brilliant theoretical physicist." Murph said that Hans Bethe once walked into a Jason meeting and said, "This looks like the Who's Who of American physics."

In spite of being pure researchers and mostly theorists, they thought the technical problems were fun and made them feel useful. "I found the work interesting," Treiman said. "It was particularily so for somebody like myself whose academic work tended to be abstract. It was nice to find that you could solve more practical problems. You could show off with your fancy techniques, you see, and dazzle people." Another Jason, Edwin Salpeter, an astrophysicist at Cornell, said he wanted to prove to himself that he could be intellectually broad, "a grimy dirty engineer as well as an esoteric theorist." And besides, he said, he had the feeling that by working with the government, he "could do something towards at least preventing a genuine holocaust." Fitch said, "Inventing and advising is what physicists love to do."

A new member in 1966 was Richard Garwin. Garwin had worked on the hydrogen bomb at Los Alamos with Teller, had been at Chicago with Murph, and then in 1952 had moved from academia to industry, to IBM, where he remained for the next forty years. He had briefed Jason during its first summers on missile defense problems—"these people," he said, "they need not only smarts but they need facts"—which he had learned about while serving on a PSAC panel. He joined, he said, because "Jason seemed to be a useful thing. Kind of family. Bunch of nice people."

CHAPTER FOUR

Heroes

"*I often encountered groups that would say, 'Why don't you god-damn scientists just stop doing these things?' And I pointed out that scientists as a group are just as patriotic as the next person. Individuals can act on their own consciences as to whether they want to participate in that or not. But to expect the scientists to withhold their favors, like the women of Lysistrata, is unreasonable.*"

—Marvin (Murph) Goldberger, interview, 1999

In 1961, while the Jasons were setting off on their glory years with charged particle beams and the Hanbury Brown-Twiss technique, the United States began increasing its military presence in South Vietnam, sending in thousands of "advisers." The point was to prevent Southeast Asia from falling to the Communist empire like dominoes.

To the public, the domino theory seemed a little abstract: the Communist menace in Southeast Asia was not Nazis in Europe with a fission bomb, or the Soviets with a fusion bomb and intercontinental missiles; it was only Soviet leader Nikita Khrushchev banging his shoe on a UN podium in support of widening the area behind the iron curtain via wars of national liberation—that is, using war to liberate non-Communist countries. In this particular war, Communist North Vietnam was liberating South Vietnam, sending troops and materials south along a network of trails generally called the Ho Chi Minh Trail. To cut off the supply chain, the United States began, in early 1965, routinely bombing North Vietnam in an operation called Rolling Thunder. Shortly thereafter the United States sent ground troops to South Vietnam. For the next year the bombing escalated, as did the number of troops, and North Vietnam showed no signs of giving up or even wanting to talk about it. The war looked as though it might go on forever.

The result in America was another escalation: of public irritation, distrust, touchiness, shame, and outright anger, a growing multiform discontent. The Jasons were as unhappy about the war as anyone else. Henry Kendall, who was at MIT and was one of the early Jasons, thought that "the United States' participation in that effort was going to end badly," and that in general the Jasons nearly unanimously felt that "things were bent out of shape in Vietnam." So the Jasons who had been so useful in solving the Cold War's technical problems decided, unasked by ARPA or anyone else, to see what they could do about Vietnam.

The summer of 1964 in La Jolla, William Nierenberg—a Manhattan Project physicist who was now at the Berkeley national lab and who joined Jason in 1962—led a study on Vietnam. The exact nature of this study is a little mysterious, but it was probably on the methods of guerrilla warfare. North Vietnamese guerrillas were running an insurgency, a kind of unconventional, hit-and-run, sociological/military warfare. They were also taking advantage of Vietnam's history of unsuccessful foreign occupation—by the Chinese, then by the French, and now by the Americans—which, not surprisingly, left the Vietnamese with a strong sense of national identity. So though the South Vietnamese disliked the North Vietnamese insurgents, they also disliked their own authoritarian, American-backed government, and they disliked the American military even more. The North Vietnamese insurgents encouraged these conflicting dislikes with persuasion, torture, and terror.

Countering insurgency relies not only on military science but also on social science—an area not typically undertaken by physicists. Jason briefly got into the social science of the insurgency anyway because Murray Gell-Mann, whose interests were catholic and compelling, was interested in human behavior. "To some of his Jason colleagues," Gell-Mann's biographer wrote, "Gell-Mann seemed far less interested in fighting the war than in understanding its sociology. He didn't see why Jason shouldn't be able to get together a group of experts in many different fields and figure out a solution." So for the 1964 summer study on Vietnam, Gell-Mann invited one of these experts as briefer: a professor of international relations who also wrote popular books and articles named Bernard Fall. That summer Fall gave Jasons lectures on the history and culture of Vietnam. After listening to Fall and other briefers, said Hal Lewis, "we came out knowing more about sociology of Vietnam than of Georgia." Nierenberg said, "We were briefed to hell and gone."

Nierenberg's reaction to the briefings on the insurgency was strong: "Some of the briefers' ideas were disgusting, I don't like to say them to a lady. But they were disgusting and stupid and idiotic." After the briefings the Jasons wrote two reports: one, called "Night Vision for Counterinsurgents," was presumably a technical paper. The other, which had a large number of authors, was called "Working Paper on Internal Warfare" and must have been a more general response to the insurgents' techniques. The latter report "had little direct impact," Nierenberg wrote in a history of Jason and the Vietnam War. But its "educational shock was so great that, by common agreement, no systematic work was done on the subject in 1965."

In 1965 Jason met on the East Coast again, probably at what was then Otis Air Force Base near Falmouth, Massachusetts, and went back to ballistic missile defense. Its reports had titles like "Whistlers as a Launch Phase Early Warning System" and "RF Breakdown Near a Conical Re-Entry Body." The summer after that, however, a third of the reports Jason produced were on Vietnam.

In January 1966 Gell-Mann pushed for Jason to study Vietnam again; and Murph, who was still chairman, agreed. By the spring meeting in April 1966 Jasons decided they could be most useful if they found a way to cut the North Vietnamese supply route, the Ho Chi Minh Trail, by means other than Rolling Thunder. Jasons put it on the agenda for their summer study, this year set for the West Coast, in Santa Barbara.

At the same time, Murph said, "quite independently of us, the pundits in Cambridge came to a similar, unrelated conclusion." The pundits in Cambridge were, of course, the well-connected, old-guard, MIT/Harvard advisers, the River Charles crowd. They were calling themselves the Cambridge Discussion Group and, concerned about the conduct of the war, had been meeting every few weeks; the group included Jerome Wiesner (who had been President Kennedy's science adviser), Jerrold Zacharias, and George Kistiakowsky (who as Eisenhower's science adviser had gone to Jason's first meeting). "The old-time warriors decided with the characteristic modesty of physicists," Murph said, "that they ought to get into this and clean it up." That same January the Cambridge Discussion Group also decided the Ho Chi Minh Trail should be cut; they got the Defense Department's approval, and in March they too put it on the agenda for a summer study.

To set up their summer study, the Cambridge group called Jack Ruina, who was now president of IDA. "I got a call either from

George Kistiakowsky or Jerry Wiesner or one of those guys," Ruina said. "Zacharias maybe. So what were they talking? They said, 'We would like to have a study on a Vietnam issue and would you be willing to set up a study so it would be an IDA study?'" IDA, with its academic trustees and its highly placed Defense Department customers, was a natural for the Cambridge group. Ruina thought it was a good idea. "And I agreed very quickly and readily," he said. "The project was not accountable to me—we almost were the mail drop for the project. We handled the administrative things."

One of the administrative things was where to have the study. The group couldn't meet on a university campus because their work would be highly classified and "in those days classified work on campuses was a controversial subject, people were nervous about doing that," said Ruina. "And I don't know how it came about that it should be held at a girls' school in Wellesley." The school was called Dana Hall and had the advantage of being empty in the summer. "And then the question came about," said Ruina, "suddenly at this girls' school you're going to see guards and safes and military guys, what will we say it is?" Ruina had the bright idea of bypassing explanations by saying it was a Jason study. Jason was scheduled to meet in Santa Barbara the next summer, and Ruina said, "Well, we run Jason. We'll say this year we're having a Jason project in the West and we're also having a Jason project in the East." So the Cambridge Discussion Group got the administrative name of Jason East, Ruina said: "I made it up as a cover."

Some of the people invited to join the Cambridge studies at Dana Hall also happened to be regular Jasons. And because the Cambridge group and Jason were, as Ruina said, "fiddling around with related subjects"—that is, cutting the Ho Chi Minh Trail— they decided to merge or at least coordinate closely.

The following June 13 the Dana Hall study met for a summer study's usual two weeks, until June 25. The briefers "were impressive and at the highest level," Nierenberg wrote, "and in a very short time, the group was up to speed." The scientists sat around a table on the grounds, "one fine afternoon," said Seymour Deitchman, an IDA engineer with experience in Vietnam who was working with Jason; they talked about sensors and aircraft and electronics, and "sketched out the general outlines of an electronic barrier system." Afterward Kistiakowsky sent a letter to the secretary of defense, Robert McNamara, that began modestly: "Dear Mr. Secretary: The eight days of briefings certainly have not made us into Vietnam experts or enabled us to reach well-founded conclusions." It nevertheless went on to be a tough letter, saying that the scientists were "forcibly impressed by the extraordinary unreliability and uncertainty of [the military's] data," and that according to the briefers, the bombing was neither damaging North Vietnam nor affecting the infiltration south. It suggested that "choke points" be found and that "interdictory force fields" be set up, "although, generally speaking, we do not propose to become involved in a broad effort at inventing new gadgets." It was signed, "George Kistiakowsky for Jason East."

The scientists at the Dana Hall meeting laid out four studies, based on the problems Kistiakowsky's letter had outlined, on which to spend the rest of the summer. The first and most famous, to be done by Jason, was cutting the Ho Chi Minh Trail with a barrier of sensors. The second study, to be done by IDA, was on the reliability of the military's data on North Vietnamese casualties and on halting infiltration: "factual information," Kistiakowsky wrote later, "was essentially nil." The third, to be done by MIT's Lincoln Laboratory, was a vague study on electronics; Kistiakowsky's letter had suggested "selective electronic jamming" of the enemy's communications. The fourth—by several Jasons, the Cambridge group, and IDA staff—was an initial study on the effectiveness of the bombing campaign, Rolling Thunder; a more detailed study followed.

Whether the studies on data reliability and electronics had any impact is unclear. What is clear is that the Jason/Cambridge/IDA studies on Rolling Thunder played a part in Robert McNamara's eventual resignation, and that the Jason study on the sensor barrier became the prototype for the modern electronic battlefield and arguably changed the way war is waged.

Rolling Thunder had begun in February 1965. Its goal was to scare the North Vietnamese into negotiations, knock out their fuel supply, and stop them from sending the troops and supplies down the Ho Chi Minh Trail. Toward that end Rolling Thunder aircraft bombed power plants, bridges, fuel storage depots, roads, manufacturing plants, railroads, ports, trains, and barges. By the summer of 1965, according to the classified Defense Department history of the war later called *The Pentagon Papers*, the number of sorties flown had quadrupled or quintupled; and in 1966 the number tripled again. The question was whether it was doing any good.

The Jason/Cambridge/IDA report, *The Effects of U.S. Bombing in North Vietnam,* was finished by August 1966; it "did not mince words or fudge its conclusions," *The Pentagon Papers* said later, "but stated them bluntly and forcefully." The report's summary began with a sentence that is still quoted: "As of July 1966 the U.S. bombing of North Vietnam (NVN) has had no measurable direct effect on Hanoi's ability to mount and support military operations in the South at the current level." North Vietnam was unaffected by the bombing because its economy was based on subsistence farming, which is hard to bomb; its more bombable industries weren't producing anything military; and whatever supplies its military needed were funded by China and the Soviet Union. In the year and a half since the bombing began, the infiltration of North Vietnamese troops and supplies into South Vietnam had only accelerated. The U.S. military had underestimated "the tenacity and recuperative capabilities of the North Vietnamese," said the report, and had failed to understand that bombing a society only makes it more united, more resilient, and more determined to resist. Two months later, in October 1966, Robert McNamara wrote President Lyndon Johnson a memo about the bombing, parts of which were lifted directly from the Jason/Cambridge/IDA report, including that first sentence. The enemy was just waiting us out, McNamara concluded. Rolling Thunder rolled on nevertheless: the military thought that making the war costly and difficult would give North Vietnam incentive to negotiate.

Meanwhile, a year after the 1966 study, IDA sponsored a second study on the bombing and in December 1967 sent McNamara a much longer report. It took up four volumes and was called "The Bombing of North Vietnam"; and though it wasn't a Jason report, its authors included Gell-Mann, Murph, Hal Lewis, and a new Jason, Gordon MacDonald, who was then a vice president at IDA. *The Pentagon Papers* called the report "probably the most categorical rejection of bombing as a tool of our policy in Southeast Asia to be made before or since by an official or semi-official group." The first volume summarized the state of the war; its first sentence repeated almost exactly that of the 1966 report, edited slightly for effectiveness. The next three volumes considered alternative strategies for bombing—including mining ports or attacking dikes— but concluded that none would stop the infiltration. "We looked at the best way it could be done and even that way it wouldn't do much good," said Hal Lewis. "Bombing was just not a good way to fight North Vietnam."

MacDonald, who'd spoken with McNamara about the report, thought it had a large impact on the secretary of defense. Murph thought, however, that McNamara had been convinced before the report was written: "McNamara was, to my personal knowledge, completely disillusioned about the war in Vietnam in the early summer of 1967." MacDonald and Murph are probably both right. McNamara wrote that of all the studies, private and government, on the effects of the bombing, the 1966 and 1967 Jason reports were "two of the most important," and again quoted that famous first sentence. In late November 1967, a few weeks before the report was published, McNamara resigned, unable to convince President Johnson to stop bombing. In his farewell speech the following February, according to *The Pentagon Papers*, "he drew on much of the analysis provided to him the previous fall by the Jasons."

The second idea to come out of the summer study at Dana Hall had an impact that was even larger, but completely unintended. The original Cambridge Discussion Group had included a Harvard professor of law named Roger Fisher, who had an interest in international negotiations. In January 1966, while Jason was still trying to choose a Vietnam-related summer study, Fisher sent a memo to a former Harvard colleague, John McNaughton, now the assistant secretary of defense for international security affairs, proposing to cut the Ho Chi Minh Trail with a barrier of barbed wire, mines, trenches, a swath of defoliation, and so-called strong points, essentially little forts. McNaughton liked Fisher's idea, revised it a little, and passed it on to Robert McNamara. McNamara had it sent around the Defense Department for comment by the people who would have to carry it out.

Within a few months, by March 1966, the people—the Joint Chiefs of Staff or JCS—to whom McNamara had sent Fisher's idea sent it on to the Commander in Chief of the Pacific or CINCPAC. In April 1966 CINCPAC, whose name was Admiral Ulysses Sharp, said he hated the idea: building and defending such a barrier, he wrote, would require too many troops, too much time, and too many supplies and would "deny us the military advantages of flexibility in employment of forces." Meanwhile the Cambridge Discussion Group had met with McNamara and offered to have what became the Dana Hall study look at the Fisher plan. McNamara liked that suggestion, too; and about the time that CINCPAC was objecting to the barrier, McNamara wrote back to the Cambridge group asking that their summer study examine the feasibility of a slightly fancier idea, "a 'fence' across the infiltration trails, warning systems, reconnaissance (especially night) methods, night vision devices, defoliation techniques, and area-denial weapons."

That same April Jasons, who knew none of this, had decided at their spring meeting that cutting the Ho Chi Minh Trail, compared to the wholesale and apparently useless bombing of the North, was more focused and more palatable—"the cleanest and least-killing," said Nierenberg. Fortunately two months later, before Jason reinvented the Cambridge wheel, it merged agendas with them and met at Dana Hall. The Jasons were assigned to work out the barrier's technical details. Cambridge did politics, not technology. "Kistie and the rest knew how to get things done," said Nierenberg, "but not what to do. We were the opposite." Cambridge stayed interested, though, and after the Dana Hall meeting, representatives of both Cambridge and Jason went to Washington and gave a list of subjects for further study to McNamara. McNamara said to go ahead.

"It was at this meeting," Nierenberg wrote, "that we first became aware of McNamara's deep interest in the possibilities of a mechanical barrier built of chain link fencing, barbed wire, guard towers, and a no-man's-land"—in short, the original Fisher/Cambridge barrier. McNamara told the assembled scientists that he wouldn't need any help with that physical barrier—which had apparently disappeared into the depths of the military and was being developed there—but that he'd like them to be thinking about a different kind of barrier that wouldn't require additional troops to construct or guard it.

So later that June the Jasons went to work on another kind of anti-infiltration barrier. They met on the Pacific coast, at the University of California at Santa Barbara, on the upper floor of a dormitory; after hours Sy Deitchman could look out the window and see Henry Kendall surfing. Jasons did their homework. They read up on historical barriers, what had worked and what hadn't. They studied whether current sensors could be adapted or whether they'd have to invent new ones. They learned about weaponry. Garwin, who was a brand-new Jason, continued his role as general briefer based on what he'd learned on PSAC panels: "So I gave lectures about various kinds of munitions, ordinary bombs and cluster munitions and fuzing and mines, and various things that people didn't know about because they had been working mostly in ABM and whatnot," he said. "I was just mostly a wise old man for them." At the time, he was thirty-eight. "What Garwin especially did," said Deitchman, "was tear apart the barrier design by showing all thirty-odd countermeasures the North Vietnamese could use against it."

Jasons also heard briefings about the trail. The trail was not a single thing, Nierenberg said, but "an anastomosed structure"— that is, a branching structure, like the veins in a leaf or the blood vessels in a body or the tributaries of a river; in short, it was a nice structure for the process of supply. The briefings on the trail, Nierenberg wrote, were "replete with relief maps and experts," and detailed "the main trail, nodes, rest camps, bypasses, and so on." One briefer was a geologist who had used the transcripts from prisoner interrogations—"we went by big karst cliffs in the morning, and then we crossed a river"—to painstakingly reconstruct the trail's geology. Jasons asked questions about how many people and trucks were using the main trail; about whether other trails were also used; and about which of all the trails was most effective. "I read French documents going back to 1890," Nierenberg said. "We learned a great deal about the Ho Chi Minh Trail."

The Ho Chi Minh Trail was not only anastomosing, parts of it weren't in Vietnam. On a map Vietnam and Laos look like nested arcs; at the top of the Vietnamese arc is North Vietnam, at the bottom is South Vietnam. The fastest way from North to South Vietnam is not to follow the arc but to cut straight south through Laos, and that's what part of the Ho Chi Minh Trail did, though without Laos's explicit consent. The American military had agreed, however, to not send soldiers into Laos. "We weren't allowed on the ground in Laos," said Garwin. "This whole thing had to be done from the air." Accordingly the report that Jason turned in the following August was called "The Air-Supported Anti-Infiltration Barrier." Unlike Cambridge's physical barrier, this one was nearly invisible: no trenches, no fences, no little forts.

The idea was clever, simple, almost childlike: from the air, planes would drop noisemakers down along the trails, then drop noise detectors into the trees above. Enemy troops would trigger the noisemakers, which would in turn trigger the noise detectors. The detectors—which had little transmitters—would signal an airplane circling overhead, which would relay the signal to a central computer, which would calculate the noisemakers' and therefore the enemy's position, and send that information on to the air force. The air force would go bomb whoever triggered the noisemakers in the first place.

The report began by describing two kinds of infiltration: one of trucks carrying supplies, the other of troops. The trucks used roads that ran south across the Demilitarized Zone (or DMZ, which divided the North from the South, running from the sea through the coastal plain and into the mountains along the Laotian border). The troops used two main trails, one through Laos, the other across the DMZ. The report proposed to place barriers to both truck and troop infiltration along the DMZ, where the supply system, it said, "necks down."

The report proposed noisemakers that ranged from innocuous to harmful. The innocuous ones were "aspirin-sized" button bomblets, essentially cherry bombs; Murph called them firecrackers and said they were "not to blow people's legs off, just to give a bang." The harmful ones were gravel mines, little beanbags full of plastic pellets that would explode and injure the person who stepped on them; they were partly noisemakers and partly "area denial" devices to keep people from crossing the area over which they were strewn. The beanbag part of the gravel mine could be colored to match the leaves or the clay of the trail. To complement the gravel mines, Val Fitch and Leon Lederman designed what they called pencil mines: little projectiles that looked like ballpoint pens, "air-dropped and fin-stabilized," that would burrow into the ground and blow up when someone stepped on the plunger. Like the gravel mines, they were designed only to injure. "I would say that's not one of my greater publications," said Fitch. "Or Leon's."

For troop trails, noisemakers would be laid down crossing the trail every kilometer for twenty kilometers, and then scattered over an unusually large surrounding area, a hundred kilometers long and several kilometers wide—the area in "area denial" is usually measured only in meters—to keep people from taking or making trails nearby. The noise detectors would have their own batteries and would be attached to seven-foot parachutes that hung in trees; some of them would be attached to spikes that stuck into the ground. They could hear trucks at two thousand feet and button bomblets at two hundred feet.

Orbiting over this whole system twenty-four hours a day, seven

days a week, would be an airplane that received the system's signals
and passed them on to a computer at a ground station in a village
called Nakhon Phanom, Thailand. Nakhon Phanom was the site of
a U.S. Air Force base; its military inhabitants called it NKP or
sometimes, Naked Fanny. At Nakhon Phanom a state-of-the-art
IBM 360-50 computer would analyze the signals, Garwin said, "to
try to characterize the sounds so you wouldn't be bombing birds
or peasants but convoys, trucks, or whatever." From the nonbird,
nonpeasant signals, the computer would calculate locations and,
said Garwin, order "response, immediate response." The response
was an air strike using cluster bombs called sadeyes, bombs that
blew up just above the ground, sprayed secondary explosions eight
hundred feet in all directions, and were meant to kill.

The Jason report considered geology and vegetation, listed the
habits of the infiltrators, and specified types of mines and bombs.
It suggested aircraft appropriate for dropping, orbiting, and striking. It recommended ground reconnaissance to check for effectiveness and suggested time intervals for reseeding the noisemakers. It
worried about battery life and about whether the noisemakers
would get caught in the trees or would need plastic liners to protect against damp.

"Considerable cleverness" is needed, the Jasons wrote, "since
we are pitting a technical system against highly determined and
ingenious human beings on the ground." Accordingly, the report
said, "after some period of time which we cannot estimate, but
which we fear may be short," the enemy would find countermeasures: mine-sweeping, what the report called "sensor spoof-
ing," antiaircraft fire, and route-changing. The report proposed
counter-countermeasures and envisaged, it said, "a dynamic 'battle
of the barrier.'" Jasons figured it would cost $800 million per year,
mostly for mines and bombs; they proposed a "tri-Service" task
force to further plan the barrier, improve its components, and carry

it out. They didn't think the barrier would stop all the infiltration or make the enemy surrender. But "if real constraints could be put on the ability of North Vietnam to move troops south," the report said, "it may then become possible gradually to reduce the scale of the fighting," and that would in turn, it hoped, allow "the war to taper off."

"Our objective was not to kill the North Vietnamese," said Murph, "but to lower the temperature of the war so it could be solved by political means."

"Motives," said Hal Lewis, "were as pure as the driven snow."

Toward the end of the summer Kistiakowsky came out to Santa Barbara to see what Jason had done. Then he called a "plenary session" of Jason and the Cambridge group, the week of August 15, 1966, again at Dana Hall where they signed off on a summary of the report. On August 30 Nierenberg, Deitchman, Kistiakowsky, Ruina, Jerome Wiesner, and Jerrold Zacharias met with Robert McNamara and presented their report. McNamara interrupted their presentation, Deitchman said: "He said, 'I read your report, I don't need any briefing on it but I have some questions,' and then he turned to pages that he'd tabbed." Ruina said, "The issue that McNamara was hammering on was, stop the bombing. He didn't say, 'Win the war.' He said, 'The bombing is not doing us any good, we're wasting a lot of weaponry, and we've got to find a way to stop the bombing.'"

On September 3 McNamara wrote to the JCS, the Joint Chiefs, requesting that they let him know what they thought about the air-supported anti-infiltration barrier; JCS forwarded the plan to the Pacific commander, CINCPAC, with the same request. On September 7 McNamara helicoptered in to meet with Jason and the Cambridge group for the last time, at Zacharias's summer house on Cape Cod. Deitchman and Kistiakowsky explained the plan to

McNamara; maps of Southeast Asia were spread out on the living room floor, a Zacharias dog kept crossing the DMZ, drinks and food were served. "It was, you know, a typical social occasion," said MacDonald, except that "you just were deciding the next years of the Vietnam War." McNamara asked whether the North Vietnamese were likely to use the countermeasures Garwin thought up; Deitchman said the North Vietnamese were more likely to "just try to bull through." Deitchman thought this assessment "helped McNamara decide to go ahead."

Meanwhile the military had gotten back to McNamara with its opinion of the anti-infiltration barrier: JCS was reserved, CINCPAC hostile. On September 15 McNamara again overruled their objections and ordered the barrier into reality. Even the Jasons were surprised. "When we proposed the barrier," said Lewis, "we actually proposed further study, you know, the way physicists do. Instead he created the project, gave it high priority, and jammed it down throats of military." Garwin said, "McNamara just implemented it. He gave it to Starbird and said, 'Do this.' " "This" was the tri-service task force that Jason had recommended to take charge of the barrier, and Lieutenant General Alfred Dodd Starbird set it up.

The task force was called the Defense Communications Planning Group, or DCPG, a name chosen for its meaninglessness. As the Jason report had suggested, its membership included officers from all three services, air force, army, and navy. Its adjunct Scientific Advisory Committee included, at one time or another, a reasonable fraction of Jasons: Richard Garwin, Murph Goldberger, Val Fitch, Gordon MacDonald, Henry Kendall, Charles Townes, Bill Nierenberg, Hal Lewis, and probably others; Kistiakowsky was the committee's chairman. It reported directly to McNamara. It had "unlimited acquisition freedom" and was considered "an entity outside traditional channels." The deadline for having the barrier up and operating was one year away.

At this point, somewhere in the still-classified interior of DCPG, Jason's air-supported anti-infiltration barrier was grafted onto the Fisher/Cambridge physical barrier. The Fisher/Cambridge part of the barrier was eventually called McNamara's Fence, also McNamara's Line and McNamara's Wall. The whole barrier system was code-named Practice Nine.

On October 14 McNamara, just back from Vietnam, wrote a memo to President Johnson recommending, among other things, that the bombing be leveled off and that "an infiltration barrier" be installed. The cost was to be $1 billion. The JCS wrote an immediate countermemo, disagreeing with almost everything McNamara recommended—they very much wanted to continue bombing—but agreeing in principle with the barrier if it didn't divert too many resources. Within a month the third in the military chain of command, the Vietnam command, called the Military Assistance Command or MACV, weighed in. MACV's commander, General William Westmoreland, wrote to CINCPAC saying he had "on no occasion recommended nor concurred in a barrier undertaking." Once again McNamara overruled them both and by the end of 1966 had their involuntary cooperation. On January 12, 1967, President Johnson declared Practice Nine the "highest national priority."

The Practice Nine barrier was to be built in two parts and would look for all the world like the Jason and the Cambridge barriers strung end to end. The Cambridge part of the barrier would run from the sea, west across the coastal plain, and up to the mountains; it was to be a physical barrier of barbed wire, minefields, infrared detectors, searchlights, radars, guard towers, and strong points. The Jason part, the air-supported barrier, would run from the western end of the physical barrier, through the mountains, and into Laos; it was itself two barriers, as the Jason report had dictated, an antitroop barrier and an antitruck barrier.

For the air-supported barrier, the DCPG task force ordered that the noisemakers and noise detectors be improved. The greatest improvements were in the detectors, called sensors. The noise sensors, versions of the sonobuoys used under water in antisubmarine warfare, had been modified to work in air and were called, in deference to their naval antecedents, acoubuoys. Most new sensors, however, didn't detect noise. Seismic sensors, called spikebuoys or SIDs (seismic intrusion detectors), could detect vibrations in the ground set off by trucks: a SID protected with rocket casing would drop from a plane and lodge itself into the ground all the way up to its detachable afterbody, or DAFT, which held the antenna. A fancy combination of sound and seismic sensors was called an acousid. Later in the war magnetic sensors could detect rifles and tanks; infrared sensors detected heat and therefore warm bodies and trucks; chemical sensors detected the ammonia in human sweat and urine.

The problem with most of the sensors, of course, was in knowing what they sensed. Spikebuoys and acoubuoys would detect not only convoys of trucks and battalions of troops, but also elephants, earthquakes, bombs, helicopters, rain, wind, and farmers with cattle herds. But because the sensors were laid down in patterns—strings of sensors crossing a trail, the strings a kilometer apart—the signals of a convoy or battalion moving down the trail would also come in patterns. Signals from seismic sensors could also be compared with signals from sound sensors—battalions don't moo.

To work out glitches in the system, DCPG conducted field tests. Jack Ruina went "trudging through Panama jungles," with several Jasons—among them Murray Gell-Mann—to see how well the sound sensors worked, "given the wildlife and whatnot, the acoustical conditions," Ruina said. After the test, he said, "Murray decided to stay on, to go hiking through the jungles. I came back not being very impressed with this being a great breakthrough in science, technology, or warfare."

. DCPG also improved its airplanes. The slow, low-flying, two-engined antisubmarine patrol planes called P-2s, used for dropping sensors, were easily shot down and so were replaced with fighter jets or helicopters. The P-2s had also been meant to be the orbiting airplanes that would relay the sensors' signals to the computer, but the air force decided instead to use large Lockheed Constellation aircraft, call sign Batcat. Batcats carried a large crew and a backup locator-computer, orbited out of antiaircraft range at twenty thousand feet, and as Garwin said, "never made any sense at all." Batcats were eventually replaced by little single-engine Beechcraft Debonairs, which were renamed Pave Eagles; unmanned drones were considered but were probably never used. Garwin, who has a well-filed and completely accessible memory, saw some trials for landing drones, which, he said, "had very crude controls. In fact, General John Lavelle—a pilot—was a passenger in this unmanned aircraft. And never again." Did we almost lose a general? "Yeah," Garwin said, and his voice had a little smile in it.

By the summer of 1967 Jason was back to doing its usual Cold War missile re-entry projects, and Practice Nine was guided mostly by DCPG. On June 13 DCPG changed the name Practice Nine to Illinois City, and then a month later changed it again to Dye Marker. About the same time another field test like the one in Panama, this one at Eglin Air Force Base in Florida, was testing button bomblets, when a big storm came up and washed thousands of the tiny bombs onto the Florida beaches. Miles of beaches had to be swept; first the air force said the bombs were dangerous, then it said they weren't. For this and other reasons, the highly classified barrier was beginning to show up in the newspapers, first as a physical barrier, then as an "electronic" one with creative variations.

Taking control of the increasing publicity, on September 7, 1967, McNamara announced the barrier in a press conference, suggesting it provided an alternative to a wider war but otherwise remaining vague: just a fifteen-mile clearing through the jungle and equipment ranging from "barbed wire to highly sophisticated devices." A few days later McNamara wrote to President Johnson, "I understand you were surprised by my comments at the Thursday, September 7, news conference on the so-called infiltration barrier in Vietnam," and went on to summarize the barrier. The fifteen-mile stretch was the physical barrier, he said; the air-supported sensor barrier was "unique." He thought the barrier would block 200 or 300 percent more trucks and 30 percent more troops. The JCS and MACV liked the plan, he told the president, better than they used to.

At the same time DCPG changed the name again: Dye Marker became the name for the physical barrier only, and Muscle Shoals became the name for the air-supported sensor barrier; Muscle Shoals was then further subdivided into Dump Truck, for the anti-troop barrier, and Mud River, for the antitruck barrier. By the end of November 1967 Mud River had been rehearsed and found worthy; in December it went into operation. At first it wasn't working well because the air force wanted to double-check the sensors with intelligence from light planes that flew out to confirm the targets—"confirmation for sake of confirmation," says a MACV history. To the air force, the sensors were "just intelligence that they would act on the next day, ho hum," Garwin said. Shortly thereafter, when an air force general—the same General Lavelle who tested the drone—took over DCPG from the army's Starbird, the response became "real time"—that is, airplanes would drop bombs minutes after the sensors went off. "That's what we were after in Vietnam," Garwin said.

Garwin and other members of DCPG's science advisory group

visited the computer control center at Nakhon Phanom. They put on earphones and "listened to North Vietnamese truck drivers," said Lewis, "talking about what their experience had been during the day—'we lost four trucks.'" Garwin said the earphones were accompanied by a television monitor, on which "you could see the sensors plotted on the screen. If the sensor was active, it changed color. And you could listen in real time, so you could see a vehicle pass from one sensor to the next." A MACV history says that the first weeks of Mud River were "extraordinarily hectic," what with a still-unproven system, along with a "flood of distinguished visitors" with "innumerable urgent messages questioning, recommending, and requesting data and explanations"—the flood of visitors sounding pretty much like Garwin in action.

According to the MACV history, the antitruck Mud River worked. At the end of the month, the antitroop barrier, Dump Truck, was supposed to be tested but wasn't. Instead, the resources allotted to Dump Truck were diverted, on January 19, 1968, to the siege at Khe Sanh.

Khe Sanh was a Marine base on a plateau, near a village of the same name. It was situated south of the DMZ in the mountains along the Laotian border, not so far from where Dump Truck was to have run. To Khe Sanh's east, north, and west were high hills, which the Marines also held; and on the hills beyond those hills were the North Vietnamese. This had been the situation for roughly a year, but by mid-January 1968 intelligence reports showed the North Vietnamese adding enormously to their troops in the hills. On January 31 the North Vietnamese began—there and all along the DMZ—the massive invasion of the South called the Tet Offensive. By this time the North Vietnamese were not only in the hills surrounding Khe Sanh, they had also captured the road to the south, Route 9, that led out through hilltop passes to the coastal plain. Khe Sanh was effectively cut off; the only way in and out was by air and under fire.

The fear, both there and in Washington, was that Khe Sanh would be a repeat of the 1954 French disaster at another sur-rounded, besieged Vietnamese base, Dien Bien Phu. At Dien Bien Phu the North Vietnamese overran the outposts, then by digging tunnels and trenches, gradually moved in toward the base, firing antiaircraft guns and artillery, until the French ran out of weapons, food, and men, and the base fell. The situation at Khe Sanh was remarkably similar; in fact, the siege was being conducted by the same North Vietnamese general. "The North Vietnamese were digging under the lines at Khe Sanh," said Nierenberg, "sapping, bringing the siege in tighter, cutting off more and more supplies." Gordon MacDonald visited Khe Sanh during the siege: "It was a scary place. Because you knew you were isolated. There were something on the order of four thousand Marines and to many, there was very little hope of getting them out. It was a dreadful situation."

The number of Marines was actually somewhere between 5,000 and 6,000, and the number of the North Vietnamese was around 20,000; the siege lasted over seventy days. The enemy was mostly in-visible: they were most active at night, most days were foggy, and the elephant grass covering the plateau was up to twenty feet high. Within a week after MACV had ordered the sensor system into Khe Sanh, it was operating.

A Marine captain and intelligence officer named Mirza Munir Baig—who had briefed the Jasons in the summer of 1966 and who became an expert on the sensors—wrote a letter to his command-ing officer describing the nights of February 3, 4, and 5, 1968. On the night of February 3 the sensors on the ridge behind one of the western Marine-held hills went off in such a way as to indicate that the enemy might be moving toward the hill. Baig didn't quite know how to interpret the sensors, and so, he wrote, "I embraced the theory that it would be folly to become erudite in one's ignorance: Therefore, believe that which the sensors purport to say." So he believed that two battalions had reached a point near the hill, but day came and nothing further happened. The next night "more and more puzzling things occurred." Sensors went off farther down the same ridge, then went off again on the same place as the night before, then went off a third time on the same ridge but to the south, then went quiet there, then went off farther up where they'd gone off earlier, then everything went quiet. The question was, how many battalions were moving and in what directions: "am I looking at a porterborne logistic train or should we rush about, shouting, 'Aux armes'?" He waited. He figured that either the battalions were moving away from the areas with sensors or they were moving toward the Marines' hill. Deciding to believe the latter, he calculated, given where the enemy had started, what direction along the ridge they'd been moving, how fast, and where they'd be in a half-hour. He ordered, he wrote, that "we should place a target block 500×300 square meters about the anticipated enemy position, fire 350 rounds of light and medium artillery into it, and copper bottom our bet by adding a 1000 meter curtain at each end." When all this had been done, however, he couldn't confirm that they'd hit anything and concluded that the enemy had aborted the attack. As it happened, he'd forgotten one set of triggered sensors, and that forgotten battalion attacked; it "met a revolting fate," he wrote. The computer center at Nakhon Phanom recorded hundreds of enemy voices running panicked through darkness and heavy fog. A month later, on the night of February 29, the enemy began a larger attack directly on the Marine base itself. That attack, Baig wrote, "was signaled in precisely the same way as had been the hill attacks, and we were ready for them."

Throughout the long siege the use of the sensors evolved. Instead of firing at the location of every signal, the Marines learned to do what Baig did—wait until the pattern of signals indicated that the enemy had assembled and were about to attack. Such assemblies, said a MACV history, were "an extremely profitable target for pre-planned massed fire." DCPG kept up its improvements, and by mid-March the computer center at Nakhon Phanom could be partially bypassed; soldiers wore portable monitors that beeped when a sensor triggered. Finally either massive artillery fire or massive air strikes—reports vary—discouraged the North Vietnamese enough that they gradually stopped attacking and the Marines went out and took their hills back. By April 9, Route 9 out to the coastal plain was opened. At the end of June the Marines moved out of Khe Sanh, and the base was abandoned to the elephant grass.

About two years later, in mid-November 1970, a U.S. Senate Armed Services subcommittee held hearings on what they now called the electronic battlefield. The senators understood that scientific inventions like radar had previously turned out to be helpful in finding the enemy, and they wanted to know whether these electronic sensors were similarly useful. Much of the testimony was on Khe Sanh as a sort of case study for the use of the sensors; Baig's commanding officer read his Khe Sanh letter aloud. The senators asked repeatedly what good the sensors had done, how many lives they had saved, how many more enemy they killed; and the military men kept answering that they hadn't done the experiment, hadn't tried to defend Khe Sanh with and without the sensors, and so the honest answer was, they didn't know. General Rathvon McClure Tompkins, the Marines' commanding general: "I have got to say that in my professional opinion it did save American lives, and how many, of course, I don't know." Colonel David Lownds, the Marine commander: "I will go out on a limb and say,

I think the casualties would have almost doubled at Khe Sanh." Major Jerry Hudson, a Marine intelligence officer: "We were able to keep him off our back by using the sensors, disrupt his plans and inflict casualties on him." Colonel Lownds again: "If you ask me would I have liked to have been at Khe Sanh without the sensor, I would have to say no, I wouldn't."

The rest of the testimony in the electronic battlefield hearings was on the more general use of the sensors throughout Southeast Asia. In April 1968, after the siege at Khe Sanh was lifted, DCPG's latest director, Major General John Deane, had been directed to help MACV use the electronic sensors in "a wide range of tactical applications against the enemy," which Deane called a "ground tactical system." He gave the senators a definition of "ground tactical," which was simply a summary of what Baig did at Khe Sanh. He said that the sensors allowed such accurate detection of the enemy at night, in fog, behind hills, and in the jungle, that attacks on the enemy could be remote—that is, only artillery or air strikes—and would need no soldiers. With the sensors, ground war "was a new ball game," Deane told the senators. "Virtually every U.S. ground combat unit in South Vietnam is now applying sensors to detect the enemy."

The senators heard from all the services—the army, air force, and navy—all of which liked the sensors. The navy used the sensors on a shipping channel leading into Saigon: Rear Admiral William House said, "No one who used them, especially for defensive purposes, wants to be without them." Major General Deane set out the sensors on a table and demonstrated them to the senators, who were intrigued; Deane told the senators that the electronics were a little delicate, and he would "appreciate if people do not fiddle with them."

A year before the hearings, in October 1969, the MACV commander, General William Westmoreland, had given a speech to the army: "We are on the threshold of an entirely new battlefield concept," he said. "I see battlefields on which we can destroy anything we locate through instant communications and the almost instantaneous application of highly lethal firepower." During the hearings in 1970 Senator Barry Goldwater, who had visited Nakhon Phanom, said, "I personally think it has the possibility of being one of the greatest steps forward in warfare since gunpowder."

The one thing the sensors didn't do was what the Jasons had wanted them to do in the first place: lower the temperature of the war by slowing infiltration down the Ho Chi Minh Trail. In the first place, the antitroop part of the barrier, Dump Truck, never did get set up. Once the siege at Khe Sanh was lifted, sensors were placed west of the physical barrier in the mountains, in the area that had been intended for the antitroop barrier. But the military found these sensors most useful for finding and bombing not troops but truck stops. In the second place, the antitruck part of the barrier, Mud River, didn't seem to have done much good. Trucks were bombed, but the enemy was using well-known roads and driving with the headlights on, so no fancy sensors were needed to find them. Over the next few years the sensor system— as of May 31, 1968, called again by one name, Igloo White—was used in Laos against trucks. To assess success, the military kept data including the tonnage of supplies sent from the North versus the tonnage arriving in the South. The military claimed to have reduced the supplies by 80 percent. But the data were raw numbers, and like most data then coming out of Vietnam, they were misleading, if not outright lies. Even if they'd been true, however, it didn't matter. The North Vietnamese turned out to be able to operate successfully with 20 percent of the supplies being sent, and they just kept on attacking the South.

Other reasons for the barrier's failure vary, but they come down to the air force refusing to use it as a barrier. Gordon MacDonald was on the DCPG science advisory group: "As the war went on, the air force said they had higher priorities in North and South Vietnam. Aircraft could be put to better use bombing other places than cutting off supplies." Richard Garwin was on DCPG's science advisory group: "The air force didn't like to have a kind of semiautomatic system that would tell them where to go and bomb. They don't like being told what to do, any more than Jason likes being told what to do." Sy Deitchman was on DCPG's science advisory group: "The air force's commanding general told me that he was able to engage only about fifteen percent of the targets. So it became mainly a surveillance/reconnaissance system with occasional strikes." An air force historian summed up: "While building a barrier of electrons appealed to the whiz-kids," he wrote, the anti-infiltration barrier "did little to affect the outcome of the war."

Meanwhile back home a lot of the scientists, who knew exactly how the military was using their barrier, were bailing out of DCPG. In January 1968, just as the sensors were reassigned from an anti-infiltration barrier to the ground war at Khe Sanh, Kistiakowsky resigned from DCPG's science advisory group. Charles Townes replaced Kistiakowsky as the group's head briefly, then resigned himself because it became clear to him, he said, that the military "simply didn't want this McNamara Wall." Other DCPG science advisers did the same. Murph said, "Whereas we had gone into this with the notion of lowering the temperature of the war by slowing the infiltration, that's not the way they"—meaning the military—"looked at it at all. We looked at it as a replacement of the bombing in the North, which we knew wasn't worth a damn. But they looked on us as an *add-on*." Murph said "add-on" so forcefully that his voice cracked. "They would bomb the North, *and* they would do this. So I quit after a while, I quit this DCPG."

Garwin did not quit DCPG. "One needs to separate what Jason

does and why Jason does it," he said. "Even though the motivation of the Vietnam study may have been to stop the bombing in the North, the deliverable was an objective analysis." The "deliverable" was the 1966 report "The Air-Supported Anti-Infiltration Barrier"; and its objective analysis was handed over to the military. What he was saying was, an analysis handed to a sponsor is the sponsor's to use as needed. "Now it may be that Murph quit the DCPG as a result, but I did not," Garwin continued. "The deployment of sensors around Khe Sanh was hardly optional."

In June 1972 DCPG was disbanded because its work was done: a cease-fire was being negotiated and U.S. troops were leaving Vietnam. Henry Kendall, who had been on DCPG, said, "It was clear, somewhat after the fact that the government was not interested in the advice that was given it, and didn't intend to take it, and did not take it. I found the Jason experience deeply educating." Kendall didn't want to talk about the outcome of the barrier project, although he added, "it didn't win the war, I can tell you."

The concept of the electronic barrier, however, was alive and flourishing. The director of ARPA in the early 1970s was Stephen Lukasik: for the first time since the creation of nuclear weapons and ballistic missiles, he said, we have "a totally new technology that we now call smart weapons." Even if the anti-infiltration barrier was "horribly naïve," he said, "it really gave rise to this whole notion of sensors embedded with computers." So even in the anti-infiltration barrier's "quote-failure-unquote," said Lukasik, "in a longer-term sense, it was a success. But as happens so often with discoveries, it was a success for something different."

CHAPTER FIVE

Villains

"After all, I had worked on nuclear weapons and was able to live with myself on that issue. Not that I thought that their use was warranted, but that the principles of deterrence made sense. So the question was what we as scientists could do to help. I think things were okay until Vietnam, which caused an enormous dislocation in Jason."

—Ed Frieman, interview, 2002

So the Jasons, who learned from the Manhattan Project that good science can be crucial to national security and still cause great harm, and who learned from Sputnik that giving science advice could be both useful and exciting, had given their advice during the Vietnam War and in the process had created their own genie, the electronic battlefield. Murph particularly was upset by this. "It was almost a textbook demonstration of the arrogance of physicists," he said, talking about the Cambridge group but not excluding Jason. "After all, we had won World War II, a much bigger operation. So mopping up this little war was clearly something we could do and everybody else was fucking it up. And it was the greatest mistake that any of us ever made. What we should have told McNamara at the time was to take a flying jump."

Not only did many Jasons not like what happened with their barrier, they also didn't like having been part of the war. Hal Lewis was now the head of Jason; Murph had resigned as head in 1966, only because "I just had too much on my plate," he said. "I was still trying to do research in physics, and it was enough already." Jasons "had genuine conscience issues," Lewis said. Just before the spring meeting in 1968—at about the time the sensors were helping lift the siege at Khe Sanh and just before MACV began using the sensors widely—Jasons went on a field trip to Eglin Air Force Base and sat down to discuss the war. Afterward Lewis wrote a "Memorandum for Jason Members" summarizing the Eglin discussion. The discussion included "a wide spectrum of opinions," he wrote, but "the center of gravity" was "a sense of moral outrage at the war, and a feeling that Jason ought to 'do something.'" One of those things was for Lewis to "convey the feelings of the group, without threats, to people in the highest accessible places." The next time Lewis was in Washington he went to see Paul Nitze, the deputy secretary of defense, and laid out the Jason consensus "about strategy and tactics and intent and objectives." Nitze seemed to agree with everything—"Nitze was very sympathetic," said Lewis—but the talk apparently had little effect.

Lewis's memorandum also said that Jasons had decided that their dislike of the war didn't mean they had to change their "modus operandi": anything they disliked, they could just refuse, individually or collectively, to work on. But according to Charles Townes, they did have a meeting about whether to continue working for the government. When Townes's opinion was asked, he said that Vietnamese had died and American soldiers had had "their lives at stake, and all I have at stake is my time and my friends and my reputation. I'm willing to risk that." Lewis's memorandum ended with "The subject is by no means closed."

The subject wasn't closed, not all Jasons agreed with Townes, and a number of them—maybe nine—resigned. Sam Treiman, who had been a primeval Jason, called his resignation "a quiet thing." He had "no objection about Jason, God bless them. And I didn't even resign, I just stopped coming," he said. He was even "a little ashamed about quitting," he said. "But I just felt so uncomfortable, I couldn't." Henry Kendall, an early Jason, said he quit not for any moral principle but for "the principle of effectiveness." He had just helped found a public interest group called the Union of Concerned Scientists: "By that time, I had discovered that being inside was not an effective way to change national policy."

Freeman Dyson hadn't worked on the electronic barrier even though, he said, "I thought it was reasonable for Jason to be involved. As long as I wasn't. I thought the whole war was stupid." Dyson tried resigning, but Lewis wrote and told him that the steering committee loudly rejected the possibility. Dyson wrote back to Lewis on March 18, 1968: "This leaves me in a kind of limbo, where I propose to remain until the dust settles a bit. . . . If there is some definite and useful job for me to do in Jason, I am ready and willing to do it. At the moment I do not have any such job in view."

Underneath their unhappiness seemed to be a feeling of having been betrayed, of having lost faith. And as always with lost faith, they may have felt a little silly for having had the faith in the first place. Murph said that in the late 1960s a high government official had lied to PSAC. Murph was then a PSAC member and had heard the lie. He said that his academic friends had been telling him the administration was lying, "but if you know inside information, you think everyone who is on the outside doesn't know what they're talking about. And the sad fact of it was, they knew what they were talking about and I didn't. And I've always been ashamed at how slow I was in making that realization." The Jasons in general seemed to have been slow in realizing what Garwin had said, that they were creating "deliverables" over which the creators could have no control. And sometimes those deliverables got used in ways that dismayed the creators: Murph said, "We got taken to the cleaners." The Jasons sounded as though they were coming of age.

Sid Drell didn't work on the Vietnam studies because he was concentrating on his own personal mission of helping to avoid nuclear war by, in this case, helping to limit the buildup of intercontinental missiles. "Many people who got involved in the electronic barrier went in with the best of motives, and saw some of the technical contributions they made used in ways that they feel quite unhappy about," said Drell. "But that's inevitable, you know. The laws of physics are fixed. The laws of politics change. And you're supping with the Devil in a difficult way. It's to be expected. It's unavoidable. And you have to keep your guard up."

In 1966 Jasons did one other study on Vietnam that was almost the inverse of the barrier study: from the beginning it had the problems of supping with the devil clearly in sight and seemed to have had no impact whatsoever. It was called "Tactical Nuclear Weapons in Southeast Asia" and was a true Jason study, not a Jason/Cambridge collaboration. Those collaborative studies had the impact they did partly because the Cambridge group had connections in high places and partly because those connections requested the studies and used them. These Jasons had no connections and no one asked for this study.

Jasons studied tactical nuclear weapons because they were seriously worried. "We were scared about the possible use in Vietnam," said Robert Gomer, a chemist from the University of Chicago who was probably Jason's first nonphysicist. During the 1966 spring meeting Freeman Dyson was "at some Jason party," he said, and a former chairman of the Joint Chiefs of Staff who was also close to President Johnson "just remarked in an offhand way, 'Well, it might be a good idea to throw in a nuke once in a while just to keep the other side guessing.'"

The offhand remark seems to have been made at a reception for Jason's sponsors; it reflected loose talk around the Pentagon, said Sy Deitchman at IDA, "that 'a few nukes' dropped on strategic locations, such as the Mu Gia Pass through the mountainous barrier along the North Vietnamese–Laotian border, would close that pass (and others) for good." The Mu Gia Pass was one of the points of entry to the Ho Chi Minh Trail, and though it had been bombed heavily and repeatedly, it was just as repeatedly repaired and reused. Passes like the Mu Gia with sparse populations made what Deitchman called "attractive targets" for tactical nuclear weapons. Dyson said, "Bob Gomer and I then decided that this was something we really had to take seriously."

Gomer took the initiative, Dyson said—"it wasn't me who had started it, certainly." He and Dyson were joined by Courtenay Wright, an experimental physicist also at Chicago who had been part of the Hanbury Brown-Twiss radar experiment; and Steven Weinberg, the young Jason who had been Sam Treiman's student at Princeton. Weinberg, who said he "really didn't want to participate in anything in aid of the Vietnam war effort," felt this report was "in a good cause." The four of them did the report more or less unsponsored: the money ARPA gave Jason wasn't tied tightly to a given number of reports, and Jason occasionally fit one in.

They decided that the report would be most effective if it were a technical analysis and avoided ethics. "A cry in anguish would not discourage the military," Gomer said. Weinberg thought that using tactical nuclear weapons "was a horrible idea," but that raising ethical issues "would cast doubt on the impartiality of our analysis. Anyway, whoever read our report would doubtless feel that he was as capable as we were at making ethical judgments. So the report concentrated on purely military issues." Was staying away from ethical issues difficult? "No," Dyson said flatly, meaning "of course not." The Jasons were saying not that the ethical aspects of using tactical nuclear weapons were ignorable—"scared," "anguish" and "horrible" as descriptions are closer to the ethical than to the technical—but that given the ethical motivation, analyzing the technical wisdom of using tactical nuclear weapons wasn't difficult.

The Jasons got briefings; they read published but classified war games of limited nuclear war in Southeast Asia, done by two defense think tanks. They learned that tactical nuclear weapons are small bombs like those the Livermore lab helped develop, more or less portable and ranging from fractions of a kiloton to several kilotons. Tactical weapons are also not strategic—that is, they are not meant to wipe out cities, economies, and governments; they are meant to be used on the battlefield and only on military targets. The targets for which tactical nuclear weapons are most suited are bridges, airfields, missile sites, and large masses of troops moving in dense formations. For instance, the Chinese army might be prevented from coming through the mountains, Dyson said: "you can do that wonderfully well with a few bombs."

The Jasons' report covers the existing targets, "how many weapons of what yields could be profitably expended," and what the effects of those weapons would be. The report noted that trails might make good targets if they happened to run through forests, and bombs were used to blow down all the trees. A few hundred tactical nuclear weapons, the report said, could block "all mountain passes and trails between NVN [North Vietnam] and China, or between NVN and Laos." However, the Jasons figured that a trail blocked by blown-down trees could be cleared at a rate of ten feet per man-day—a figure they said they made up—and therefore could be reopened by fifty thousand men in a month or two. "The main weakness of tree blowdown," the report said, "is that a tree can only be blown down once."

The report considered targeting trails by blocking them with a barrier of radioactive fallout. The Jasons calculated the outcome of Y megatons of fallout distributed uniformly over a rectangle of width

L and depth D, which a man is crossing at V miles per hour, T days after the explosion. They found that using tactical nuclear weapons to make a barrier continually radioactive enough to keep people from crossing it would require a thousand bombs a year. Using fewer bombs of higher yield to create the barrier "would constitute a lethal threat to a population living permanently within a distance of 200 miles on either side of it." That sentence is underlined. Such a barrier, the report says, would not be tactical but "anti-population."

The report calculated that replacing the Rolling Thunder conventional bombing campaign—at one tactical nuclear weapon for every twelve conventional bombing sorties—would require ten tactical nuclear weapons a day, or some three thousand a year. "And in spite of it all, the basic [North Vietnamese] system of supply by man-hauling and bicycle would not be destroyed," the report said, but would be "ready to spring back as soon as the nuclear bombardment would slacken."

The report then turned to what might happen if the United States used tactical nuclear weapons in southeast Asia and the Soviet Union decided to return the favor. Considering all the known Soviet and Chinese tactical nuclear weapons and all the methods available for delivering them, the report tended to favor mortars or recoilless rifles as being able to deliver the bombs most efficiently and effectively. It pointed out that the United States had thirteen main bases in South Vietnam, then offered some little scenarios of how bases might be wiped out. For instance: two U.S. trucks are stolen from a base and hidden until they're no longer looked for; one is loaded with a ten-kiloton bomb; both enter the base and stop just inside; a timer is set, the crew gets into the second truck, and is two miles away when the explosion comes. The more conventional scenario, should the Soviets so choose, would be to deliver tactical nuclear weapons by missiles: all U.S. bases were within range of various Soviet missiles based in various locales.

A chart describes the radius and type of damage from bombs of various yields exploded in the air or on the ground. A ten-kiloton groundburst at the U.S. helicopter park at An Khe would destroy all helicopters. A ten-kiloton airburst over unprotected U.S. troops 8,000 feet away from ground zero would give them second-degree burns on bare skin. Troops 3,300 feet away would get doses of a thousand rems, which are lethal. If an attack were coordinated on all U.S. targets with a hundred ten-kiloton bombs, the report said, "the U.S. fighting capability in Vietnam would be essentially annihilated." The report seemed uncomfortable with this litany: it said, "There is no need to carry this scenario further."

In short, using tactical nuclear weapons in southeast Asia was a bad idea because our side made a better target than the other side. "It was mostly just a question of how vulnerable we were," Dyson explained, "and how invulnerable the other side was." And what if those Jasons had done their technical analysis and found that using tactical nuclear weapons was a perfectly good idea? Dyson said, "We probably would have then kept quiet and not said anything." Weinberg said the Jasons' analysis was honest, "but I have to admit that its conclusions were pretty much what we expected from the beginning, and if I had not expected to reach these conclusions then, for the ethical reasons that we left out of the report, I would not have helped to write it."

The report likely went unnoticed. These four Jasons, Weinberg said, were "not strongly coupled to people at the Pentagon." Deitchman suspected the Jason report came up during one of the early meetings between Robert McNamara, the Jasons, and the Cambridge group, but only because "there was a step function," he said, meaning that before the report he'd been hearing rumors of using tactical nuclear weapons and after it he heard nothing more. But McNamara was famous for being opposed to the use of nuclear weapons, so even if he knew of the report, he wouldn't have needed it. Weinberg said that he resigned from Jason a few years later, in part "because I had no idea whether what I was doing was useful or not."

As it happened, the subject—though not the Jason report—came up again more than a year later, over the siege at Khe Sanh. The Joint Chiefs of Staff asked whether, if the situation became desperate enough, tactical nuclear weapons should be used. Over a couple of weeks in late January and early February 1968, memos flew back and forth among liaisons, deputies, the Chairman of the Joint Chiefs of Staff, the commander of the Pacific forces, the commander of the forces in Vietnam, and President Lyndon Johnson. The military men thought tactical nuclear weapons wouldn't be necessary but thought they might make contingency plans. No nuclear weapons were currently in South Vietnam, but they were on aircraft carriers and presumably nearby.

Meanwhile on Saturday, February 3, 1968, Richard Garwin, Henry Kendall, and several other scientists traveled to Vietnam to check on the operation of the electronic barrier. On Monday, February 5—the day Captain Baig was figuring out how to use the sensors at Khe Sanh—an anonymous caller suggested to a staffer for the Senate Foreign Relations Committee that the committee look into why scientists who were also nuclear weapons experts were being sent to Vietnam. The newspapers found out about the anonymous call, put the scientists' trip together with the talk around Washington, and suggested in print that the administration was considering using tactical nuclear weapons in Vietnam. The Pentagon replied that the scientists were in Vietnam but for entirely nonnuclear reasons. Only after a meeting of the Senate Foreign Relations Committee, a congressional inquiry, more news stories, and a qualified denial by the secretary of state did everyone finally believe that the administration wasn't going to use nuclear weapons in Vietnam, which indeed was the case. On February 11 Johnson had told the military to stop making contingency plans.

Garwin and Kendall—who had meanwhile been diverted to the computer headquarters at Nakhon Phanom—couldn't explain their trip because the barrier was so secret. Garwin didn't hear he'd caused a congressional inquiry until he was in the airplane over the Pacific, on the way home. "I had probably told people I was going to Vietnam, which I shouldn't have," he said. "Probably some colleagues with overheated imaginations and a sense of mission thought someone should know about this."

The only certain outcome of "Tactical Nuclear Weapons in Southeast Asia" was that the title alone—not the report and not its conclusions—was declassified. Subsequently a graduate student who had planned to study at the University of Chicago with Gomer announced publicly, Gomer said, "that he couldn't work with someone who had participated in this study."

Gomer's disapproving student and Garwin's overheated colleague were part of an increasingly ugly national mood. With each fresh escalation since the mid-1960s, the country's outlook had hardened, crystallizing into anger first against the war, then against the Johnson administration; and after that against the military, the military-industrial complex, the military-industrial-university complex; and finally against institutions in general, authority, restrictions, and rules of any kind. In the United States even now, over thirty years later, "Vietnam" is less likely to refer to the Southeast Asian country than to the mood swing in this one.

The academic world in particular was antiwar. Students and faculty picketed, protested, demonstrated, petitioned, burned draft cards, published newsletters, wrote articles, made posters, held meetings, held teach-ins, and got themselves arrested. ROTC buildings were burned down, the National Guard was called out, people were killed, strikes were called, campuses closed. One natural target for academic antiwar activists was academic scientists. Part of the activists' distrust of scientists was a neoromantic, Aquarian Age distrust of technology. Part of it was the perception that scientists were taking money as fast as they could from government, the military, and industry, which of course they were. According to the science historian Daniel Kevles, in 1964 the Defense Department funded academic science in one hundred universities; ten of those universities got $5 million each; MIT was first, with $47 million worth of contracts; Caltech at fifteenth got only $4 million which, however, was 20 percent of Caltech's budget.

Oddly enough, all this military money placed in scientists' hands went unaccompanied by scientists' approval of the war. Kevles cites numbers from a 1972 article in *Physics Today* called "Survey Finds Physicists on the Left." "Four out of five of the profession's elite," wrote Kevles, "disapproved of classified research on the campuses and opposed the administration's policies in Vietnam. Two out of three approved of the emergence of radical student activism."

The Jasons who disapproved of the war and therefore had resigned did so quietly because they didn't want to be seen to be giving in to pressure from the radical activists on their campuses. And the activists were in fact singling out Jasons. For the first and almost the only time in Jason's history, the group became fair and public game.

In April 1970 an underground newspaper called *The Student Mobilizer*—put out by the Student Mobilization Committee to End the War in Vietnam, no funding source or affiliation given— ran an article called "Counterinsurgency Research on Campus EXPOSED," in which it published excerpts from the stolen minutes of a Jason meeting.

The meeting had been part of the 1967 summer study at Falmouth Intermediate School in Falmouth, Massachusetts. The minutes were of a particular study, called the Thailand Study Group, that was a follow-up to Gell-Mann's 1964 inquiries into what the social sciences had to say about counterinsurgency and the Vietnamese. In this case, Jasons were trying to decide whether American aid might prevent Thailand from being the next domino to fall to the Communists. The Jasons brought in the usual briefers, as well as some academic social scientists. "For three weeks," wrote Sy Deitchman, who was there, "the visiting officials and social scientists spoke freely, frankly, and not always complimentarily, about Thai society, Thai government, and American activities there." Gell-Mann also wanted Jason to consider, as a complement to the physical sciences Jason, creating a social sciences Jason. All these matters were discussed but came to nothing; the physical and social scientific minds did not meet.

Sometime later a student broke into the files of an anthropologist who had been at the meeting, stole the minutes, and gave them to the Student Mobilization Committee. The *Student Mobilizer*'s article gave only a vague sense of Jason's aims, excerpts were printed out of context and in no particular order, and quotes were often unattributed and didn't make much sense. But the article left no doubt that Jason was up to nothing good. The social scientists, the sponsors, and the Jasons talked about the pros and cons of access to classified studies and of working for the government: Gell-Mann was quoted saying, "We have never stimulated a government to respond to insurgency so early and so massively—maybe we can contribute to it." When Gell-Mann asked a government sponsor interested in counterinsurgency what he wanted from the social scientists, the sponsor said, "I want tools." Gell-Mann said—no context or clarification given—"Can we find out what effect increasing police density or ear cutting or other negatives have on villager attitudes?" *The Student Mobilizer* drew a box around that quote.

Seven months later two anthropologists active in the antiwar movement who had been given Jason's stolen minutes wrote in *The New York Review of Books* that "straightforward anthropological research" was being intertwined with "overt and covert counterinsurgency activities" and that this intertwining was wrong. It noted the sponsor who wanted tools; said that the Jasons seemed dismayed about social scientists' reluctance to work for the government; and listed Jasons arguing the advantages of congeniality, money, prestige, interesting problems, and close governmental ties. Read through the filter of the times, Jason looked bad.

Eight months later, on June 13, 1971, *The New York Times* began its serial publication of *The Pentagon Papers*, and Jason suddenly looked much worse. *The Pentagon Papers* were drawn from forty-seven volumes' worth of classified history, commissioned by Robert McNamara, and called "U.S. Decision Making in Vietnam, 1945–1968." It was based on the primary documents—memos, letters, speeches, minutes—that all good histories are, and it was written by a team, one of whom was a Pentagon consultant and an analyst at RAND Corporation, a federal research center like IDA. The RAND analyst, Daniel Ellsberg, was one of the few people who had read all forty-seven volumes and who could therefore, he said, "learn the lessons of the entire sweep of that period." What he learned, he said, was that five presidents in a row had been given realistic advice about the undesirability of an "indefinitely prolonged guerrilla war" in Vietnam and that all five had ignored the advice. Ellsberg thought that if anything was going to change, "the pressure would have to come from outside the executive branch," so he photocopied all seven thousand pages of the history and leaked it. First the

Times and then *The Washington Post* published excerpts, and then everybody else did, too.

A number of pages in *The Pentagon Papers* were devoted to the Jason/Cambridge reports on the effectiveness of the bombing and on the barrier; the reports were accurately summarized, and excerpts were quoted. In the bombing reports the Jasons' rejection of Rolling Thunder made them look like heroes: "Coming as they did from a highly prestigious and respected group of policy-supporting but independent-thinking scientists and scholars," said *The Pentagon Papers*, "these views must have exercised a powerful influence on McNamara's thinking." In the barrier report *The Pentagon Papers* said that Jason "had provided the Secretary with an attractive, well-thought-out and highly detailed proposal as a real alternative to further escalation of the ineffective air war against North Vietnam." Unfortunately, the eye-catching section of those pages on Jason was the bulleted list of the barrier's mines, sensors, strike and surveillance aircraft, and cluster bombs. The only Jasons cited by name were not Jasons at all; they were members of the Cambridge group who for convenience had been called Jason East. And no distinction was made between real Jasons and Jasons-for-convenience; they were all just Jasons.

"Well," said Mildred Goldberger, "their name was mud." Jason became "the devil," said Jack Ruina. "No matter that some of them had excellent dovish credentials, they were Jason." The Jasons tell story after story. In Santa Barbara Gordon MacDonald's garage was set on fire. In Berkeley "war criminal" was painted on the street in front of Ken Watson's house. In Boulder activists broke into an office Jason was using for its summer study that year, pushed aside two Jasons blocking the door, threw typewriters on the floor, and went through the files. At Columbia University forty-nine faculty members demanded that the resident Jasons resign either from Jason or from Columbia. In New York activists demonstrated in front of

Garwin's house; once, when he was on an airplane, a young woman sitting next to him stood up and announced, "This is Dick Garwin. He is a baby killer."

In New York Murph was asked to talk to the American Physical Society about having just led the first delegation of scientists to the People's Republic of China. When demonstrators holding "war criminal" signs tried to get him to talk instead about Jason and Vietnam, Murph said, "Look. I'll either talk about China or I won't talk about anything." He offered to talk afterward about Jason and Vietnam, if they'd get a room. They did: "And I had them at a tremendous disadvantage," Murph said. "I was the lone sheriff, walking into the town and facing everybody down." He told them, "Jason made a terrible mistake. They should have told Mr. McNamara to go to hell and not have become involved at all." *The Philadelphia Inquirer* ran an article with the subheading, "Society of Physicists Torn by Moral Question": "In a quiet voice that seemed out of place in the garish East Ballroom of the New York Hilton, Dr. Marvin Goldberger . . . told 200 persons of the anguish and moral crisis he felt from his role in Jason."

What happened to Murph in New York also happened to other Jasons in Europe. Sid Drell was giving a scientific talk at the University of Rome, and students in the audience interrupted, wanting Drell "to justify American policy in Vietnam, and then they would vote to see if I would be permitted to give a seminar. And I said, 'That's an inquisition. And no way am I ever going to do that.'" Drell offered to go out with them after the seminar and said "we can spend as many hours as you want drinking and discussing the issue. Because I'm troubled too." The students refused his offer, Drell "shouted them down," and they left, then came back shortly with a much larger group and a bullhorn. "God, there were a lot of people," Drell said. "So I picked up my papers and walked right through them and walked out." Another time, at one of the summer schools where physicists talk about their own research and ask questions about others', neither Drell nor the students would back down and the whole school closed down for the summer. Still another time Drell visited Berkeley, and this time the activists took up Drell's offer: "I would give a physics seminar and then afterwards I would stay and we would discuss Vietnam. And that happened. I said, 'I'm willing to talk about Vietnam and my concerns and my worries. But I'm not going to use political correctness as an admission to give a science lecture.' I said, 'That's fascism. It's nothing but fascism. I oppose it on any forum I see, whether from the right or the left. And I will not be part of it.'"

Murray Gell-Mann had a similar experience at the Collège de France in Paris. When he insisted on giving his lecture, the college administrators escorted him out of the lecture hall; a subsequent write-up of this event by the European activists said that Gell-Mann was "bodily expelled from the Institute." Charles Townes was attacked at a scientific conference in Italy; at the time he was only in the audience, and when he raised his hand and asked the chairman for permission to respond to the activists, the chairman said no. Later Townes wrote a letter to the chairman about the "great inconsistency of allowing people to come into the meeting who were not physicists, to criticize, and not to allow me to respond"; and the chairman wrote back saying Townes was right and apologizing. "He just got emotionally off track in that meeting," Townes said. "And that was some of what was going on in the Vietnamese War—'get rid of freedom of speech.'"

A French poster called "War Professors" listed more or less accurately the names of the Jasons and read, "Physicists! Do not let the war professors speak of 'pure' physics until they have denounced their participation in Jason and condemned publicly the American war crimes." Italian physicists circulated what they called the "Trieste letter," saying that Jasons had helped develop plastic fragmentation bombs "aimed at producing cripples" and laser-guided bombs "used to destroy North Vietnamese dikes"; the letter demanded a discussion at a scientific meeting of "the 'neutrality' of science and the role of institutional science in the military-industrial complex of the big powers." Several Jasons wrote back saying that Jason had helped develop neither plastic fragmentation bombs nor laser-guided bombs. A French physicist replied, "Imagine a discussion on the chemists who advised the Nazis as to which gas to use in the gas-chambers," arguing about the distinction "between those who work on 'cyclon A'; and those who work on 'cyclon B.'" Dyson wrote, "If you sincerely want to bring the war in Viet-nam to an end, you will not waste your time and energy in disrupting scientific meetings. Such disruptions may be satisfying to your ego but they have no effect on the war."

The worst problems Jasons faced, however, were back home at two campuses, one of which was the University of California at Berkeley and centered around an activist physics professor. His name was Charles Schwartz and he was, I think, the only activist whose knowledge of Jason was personal. Schwartz had done his doctoral work in physics at MIT under the eminent Victor Weisskopf about the time that Oppenheimer was losing his clearances partly as a result of Teller's testimony. It was clear to Schwartz then, as it was to many physicists, "that Oppenheimer was the angel and Teller was the devil." Physicists' views "just became papered on to me," he said, because he wanted "to emulate and copy and to follow in their footsteps, if not overtake them." Drell offered Schwartz a postdoctoral position at Stanford: "I essentially never had to sweat . . . about even looking for a job," Schwartz said. "You know, the spoiled child." The following year he was made assistant professor but was not given tenure. He then turned back to Weisskopf: "I cornered Viki and hollered 'Get me a good job somewhere.'" He ended up at Berkeley where, in 1962, Ken Watson asked him to join the "junior Jasons," a summer study set up by Keith Brueckner, then vice president for research at IDA, Watson said, "for younger people at that time to have an introduction to science in the government." Schwartz spent one summer at the junior Jasons and found "a sense of glamor attached to it," he said, "the headiness—wow, you know, now we're really getting into it!" He worked on a study on "de-escalation in a nuclear crisis" but was not asked to return.

A few years later his brother was killed in a light plane accident, and thereafter, he said, he started getting serious and in particular, politically active. Eventually he formed a group, in 1969, called Scientists for Social and Political Action, or SESPA. SESPA's initial announcement said that scientists were increasingly financially dependent on the government and thus increasingly cautious about going public: "We must therefore strive to regain our full intellectual and political freedom." SESPA went after Jason. "Now, within Jason," Schwartz said, "were such people whom I felt could be and perhaps needed to be identified as war criminals."

In 1972 SESPA published *Science Against the People: The Story of Jason*. It begins with a history of the barrier project, based on *The Pentagon Papers* and the Senate electronic battlefield hearings. It moves on to small profiles of several Jasons—including Watson, Townes, Murph, and Drell—based on interviews Schwartz did with them about their views on working with the military; judging by the profiles, the Jasons talked openly—they saw him as a colleague, Schwartz said. Schwartz sent the small profiles back to the Jasons and got "wonderful letters in reply"; some Jasons said that he misrepresented their views and refused permission to publish. Schwartz published anyway. "That was tough on those people," he said, but "I felt they needed to be held accountable."

Science Against the People lists the Jasons more or less correctly.

Hal Lewis had written a letter refusing to give Schwartz an official list because, he said, Jasons in New York had received anonymous phone calls threatening their children. Schwartz wrote Lewis back, "We find it absurd to compare these actions on the part of a few frustrated and powerless people to the bombing, burning, maiming, and killing of millions of Asian people, which has been deliberately facilitated by the privileged Jason scientists who hide behind a veil of 'scientific objectivity' and military secrecy." *Science Against the People* ends with a demand that Jasons cease all services to the military and reveal the classified information they hold. Except for its two sources, *Science Against the People* has little information about Jason; nevertheless it is adamant in its point of view and the Jasons are depicted as foolish, if not corrupt, pawns of a militaristic government.

The Jasons don't seem to bear Schwartz much ill will. "He fought passionately about Vietnam and about scientists not working on these things," said Drell, Schwartz's old mentor. "I felt that sometimes he was more heat than light. He thought the same of me, I'm sure."

"Charlie was so extreme that it was very easy to disprove him and people saw that," said Townes, Schwartz's colleague at Berkeley. "So he wasn't as effective as he might have been."

"He became—you know the word *meshugah*?" said Murph. "He always appeared to be acting from the highest principles, but he became obsessive about it."

"Charlie Schwartz hates Jason," said Lewis, and Jason accepts it "as part of life."

The second campus on which Jason faced problems was Columbia University. Every Wednesday SESPA's New York chapter demonstrated against the Columbia Jasons, usually at Pupin Laboratories, which housed the physics department, and sometimes at the Ja-

sons' homes. Beginning in March 1973 and continuing every week
for a year, then irregularly, SESPA distributed leaflets. They were
usually titled "To Our Colleagues in Pupin" and ended with "Ask
them to leave Jason." They were mostly unsigned; those that were,
were often signed by a husband-wife team. The contents were out-
raged squawks: Jasons are "Baby Killers for Pharaoh," and the
"well-spring of their own work . . . is, simply, a quiet delight in
killing." One leaflet recounted an overheard story about a dinner
party attended by Richard Garwin. The dinner party was discussing the levels of strontium 90 from nuclear tests in the milk
that children drink, and according to the leaflet, Garwin not only
spoke in favor of atmospheric testing but added, "Well, what's a
few dead babies or mothers?" Garwin said the story wasn't true,
and in any case he supported a ban on testing; SESPA then sent out
another leaflet with another version of the story, this time saying
that Garwin had, at that dinner party, "compared babies and
mothers to the Jews in Germany."

SESPA named Jasons and published the home addresses of
three of them. One activist wrote Henry Foley, a Columbia Jason, a
letter: "You are the worst kind of war criminal, fit only to be executed, as your rotten, amoral vicious ilk, the Nazi officers, were."
Foley wrote back telling her to ask for proof before deciding what
to believe. Leon Lederman, another Jason at Columbia, wrote his
own three-page, single-spaced flyer titled "SESPA Kills, or Jason
Fights Back": "SESPA has an advantage in that its material is suc-
cinct, triple-spaced, easy to read. The Jason story is longer but please
bear with us." Lederman's flyer outlined Jason's history as well as the
barrier's history. It said that a result of Jason's work on the test
ban was "a reversal in the rapid increase of strontium 90 in the
milk drunk by children in prams all the way from Kuala Lumpur to
the entrance to Pupin." It ended by saying that SESPA's tactics—
"an admirable advance on the tactics of [Joe] McCarthy: lies,

half-truths, guilt by association, threats, invective, out-of-context quotes, rhetoric"—were damaging the causes of peace, disarmament, and demilitarization.

On April 24, 1972, the protest got physical. A group of maybe fifty people, identifying themselves as the New York Anti-War Faculty and as professors from area universities, gathered outside Low Library to protest Columbia's Jasons. Columbia's president went out to talk to them, arguing that he had to protect not only their right to protest but also the Jasons' right to associate as they saw fit. The group walked across the wide, old-world plaza in front of the library and over to Pupin, chained its doors shut, and linked arms in front of the building. They sent messages to the president asking to be arrested. The president, who thought they expected media coverage, decided that he had other problems on campus and would let the group stay where it was.

This went on for three days. One of the physicists caught inside Pupin was a postdoctoral student named William Happer. He said that the activists had taken control of the lobby and first floor before the physicists realized what had happened. "A bunch of guys with guns and clubs," he said, locked the doors and said to everyone inside that they couldn't go out, "so if you didn't have a gun or a club, you couldn't get out." Happer, who had no connection with Jason, had a conversation with an activist:

"You're a war criminal."

"Why am I a war criminal?"

"You're working on physics."

"But all I do is measure nuclear spins and optical pumping."

"You're all war criminals in this building because two of you are in Jason."

Shortly afterward the physicists cut off the rest of the building; on the floors above they bolted the lab doors and turned off the elevators, so that the only way for activists to get to the labs,

Happer said, "was to break down the doors and fight with us. And they probably would have won, but they decided not to do that." Happer's wife handed blankets and food through the window: "There was a window out to 120th Street—there were bars so there was no way to get out—but she could hand in food." The physicists slept on the floor. They were "a motley group," Happer said. "There were people like me, young faculty members. There were machinists there, big sturdy people, it was nice to have them on our side. There were older professors."

Finally physicists outside Pupin who had been getting worried about their experiments inside decided to move. A physics student asked the activists to let him get sneakers he'd left in the building, and as he entered a flying wedge of physicists pushed into the building and took it back. Each side stood and shouted at the other. Around midnight the poet-activist Allen Ginsberg and a companion poet showed up; the companion took out a harmonium and for an hour chanted a Tibetan mantra, alternating "om" with "horror" and "shame." The next morning Columbia's administration finally agreed to have the group's leaders arrested, and the activists left.

Happer had left sometime that night. The activists were sleeping—Happer said all those "om's" became soporific—and he walked out of Pupin and headed across campus, and "suddenly out of nowhere came this little band of poor little black kids from Harlem." The little kids, hanging on Happer's arms and legs, dragged him through campus to a streetlight, under which was "this white guy haranguing some other little black kids, talking about class warfare, the coming revolution, the oppressed, the victory of the proletariat." The little kids said, "Hey, we got one." The white guy looked at Happer and said, "Who are you?" Happer drew himself to his full height and said, "I'm Dr. Happer. I'm a postdoc in the physics department.'" The white guy wasn't impressed with

postdocs: "He looked at me and sneered, and he says, 'Let him go.'"
Sometime later Happer saw the same man on the front page of *The New York Daily News*. "He was one of the people involved in this armored car robbery at Nyack—they waited for the truck with the money to stop, the guards got out, they shot them both dead, took the money, and ran."

Because of the Pupin lab protest, Happer joined Jason. He'd been asked to join earlier but, he said, "I turned it down because I was busy trying to do my own research and I knew it would be very distracting. Then I was held captive for three days at Columbia, and you know, I felt my life was in jeopardy. And you don't get over that quickly." Happer thought, if Jason "collects such slimeball enemies as this, it must be pretty good. I mean, I didn't have a dog in the fight at the time. But once it was all over, I did."

In all this pro- and anti-Jason eloquence, the emotional intensity on both sides is palpable. What the intensity was about, however, is not immediately clear. The activists' facts were usually wrong and their arguments were overly simple, not backed by evidence, and unconvincing. Jasons argued back, and the activists countered with unrelated arguments. In the December 11, 1972, issue of *Christianity and Crisis,* Garwin wrote, "What is under attack is the right of an individual, in his own time, away from his regular job, to engage in a legal activity to which some individuals are opposed." An antiwar professor wrote back that Garwin's rights came from a government that was corrupt, and that by extrapolation from the Nuremberg trials, Jasons "should be tried for war crimes." Activists and Jasons were talking past each other. No one was having a conversation.

One exception was a letter written in December 1972 to a number of Jasons and signed by French, English, and Swiss scientists, including three Nobel Prize–winners, saying that the weapons recommended by the barrier report "have caused terrible wounds among Vietnamese civilians." Jasons wrote back saying that the weapons were not new and in any case were not developed by Jason; that as horrible as those weapons were, they were far less terrible than the wholesale bombing they were intended to replace; and that if the barrier had been used as Jason intended, fewer civilians would have died.

Another exception was an article in the November 1974 issue of *Bulletin of the Atomic Scientists*, written by an English physicist and based on letters he'd written to individual Jasons "inviting them to explain how they could justify to their consciences" their defense-related work. Of those who replied, some said that scientists as citizens are obliged to support their government; some questioned that obligation when the government's policies seem reprehensible; but most found the author's question presumptuous and irritating. Sid Drell wrote back that he was convinced that a country with scientist-intellectuals on the outside and decision-makers on the inside was unhealthy, and that individual scientists could choose the extent of their government involvement. The author said the Jasons failed to question their work in light of the work's purpose and found them morally wanting. The article, though it has a clear anti-Jason bias, is also the only one that acknowledges that the electronic barrier Jason designed was not the electronic battlefield the military used. "The scientists became, to some extent, prisoners of the group they had joined," the author wrote. At what point should they have quit? "The decisions involved," he concluded, are "delicate and difficult," but he didn't explain the delicacies.

In general, if the activists had an argument, it is that Jason was using science to help the government do harm. So what should Jasons have done differently? Were they wrong to try slowing the bombing by slowing infiltration? Probably not. Were they naïve to think scientists could intervene in military problems? In hindsight, yes; the military and the scientists had different desires, so the scientists' advice was irrelevant. During most of the Manhattan Project, the scientists' and the military's desires were the same; but during the Vietnam War, the barrier scientists' desire to stop the bombing diverged from the military's desire to stop a Communist takeover; and the case was complicated by a government that wanted to do both. Given all this, not to mention the ambivalent government, were the Jasons also naïve to think the military would use their advice for the purposes they gave it? Without a doubt.

But these are pragmatic questions. Behind them are moral ones, the ones the activists seemed to be asking. That is, behind the question of whether Jasons were naïve to help the government is the moral question, were Jasons wrong to do science with a potential for harm? Walter Kohn, a Nobel Prize–winning physicist at the University of California at Santa Barbara, was never invited to join Jason; if he had been, he would have had reservations—which he said he might also have overcome. The reason for his reservations was a moral principle: the "tremendous danger of scientists being carried away by what Oppenheimer called the technical sweetness of the project," he said, and therefore ignoring the consequences of harm.

This moral principle about the dangers of technical sweetness gets cited often, but it doesn't translate well to reality. If scientists are to abide by this moral principle, they must choose research that has no possibility of causing harm, such as studying the origin of galaxies or (like most of the early Jasons), theorizing about fundamental particles. Then if the scientists see that a line of research will lead to harm—for instance, to a military technology—they must stop working on it. And if they can't see harm at the beginning of the research—as with atomic physics and atomic bombs—they must then stop research at the point at which it suggests harm. Scientists often say that pure science is neither moral nor immoral, and that they need to consider morality only when the science becomes applied—that is, becomes a militarily useful technology.

The problem is, much of science implies a technology, and much technology is what military jargon calls "dual use"—that is, usable for both civilians and military. For instance, research into an atom's perfectly regular vibration naturally suggested atomic clocks, which themselves led to global positioning systems, used by Sunday boaters and standing armies alike. The whole argument, the whole chain of logic about doing no harm, quickly reduces to the absurd: if no idea, no invention, no outcome of scientific thought, can ever be guaranteed to be militarily harmless, then scientists should give up science. "So you'd better just stop thinking," Dyson said. Dyson has several laughs, each of which seems to mean something different; this one was a snort and was wholly dismissive.

Many scientists avoid this infinitely receding moral quandary by doing only pure research, disavowing its possible technological applications, and refusing to work directly for the military. But this is absurd, too: Kohn himself, after saying he didn't want to do science with military consequences, added that he knew his position was logically inconsistent: "It's very easy to punch holes in it. So I don't want to work on this. Well, do I think we should not have an effective defense? Of course I don't think that. So, it's all right for other people to work on it as long as I don't have to work on it? That's not very logical."

Still, in the everyday world of work, scientists do draw a line between what they will and will not work on. The easiest and most fundamental decision about the proper placement of the line is, it's individual: as Drell had said, scientists must decide for themselves.

Perhaps at the conjunction of science and the military, no universal moral law exists that says, this side of this line, good; this side, bad.

Jason has made no official decision about that line, except that individual Jasons are—as they recognized during the Vietnam studies—free to work on a problem or not, depending on their own decisions about good and bad. Individual Jasons say they wouldn't want to design new and smaller nuclear bombs, or weapons for targeted assassinations, or chemical weapons. Jasons in general agree that, as Happer said, "we would not work on any programs that went after the civilian population." Murph said Jasons preferred to work on problems of military defense, not offense, but then quickly ran aground trying to differentiate between weapons of defense and weapons of offense, which are, in practice, often inseparable.

One early Jason—Edwin Salpeter, an astrophysicist at Cornell—worked out a personal balance between (a), the "moral complicity" of working with the military, and (b), the possibility that "honest technological work" could have a desirable effect. The balance between (a) and (b), he said, changes with time. During the early 1960s, when the military was deciding between the strategies of nuclear first strike and nuclear deterrences, technical advice helped push the decision toward deterrence, and the balance for Salpeter tipped toward (b), the desirable. By the spring of 1968, however, the military was no longer taking advice, the balance tipped toward (a), the complicit, and Salpeter dropped out of Jason.

Incidentally, the same balance was being calculated by the Jasons who wrote "Tactical Nuclear Weapons in Southeast Asia." They were tipping toward (b) when they decided that technical reports can be a good medium for moral aims.

Meanwhile, the war had wound down. In 1968 Lyndon Johnson didn't want to run again, and Richard Nixon had become president. Nixon had continued bombing North Vietnam, then had widened the war by bombing and sending troops into Cambodia. Nixon's strategies worked no better than Johnson's had, so gradually Nixon reduced the number of U.S. troops; and finally in January 1973 the United States and North Vietnam agreed to stop fighting. Two years later North Vietnam captured South Vietnam's capital, Saigon, and the war was over.

Henry Abarbanel, a physicist at the University of California at San Diego, joined Jason in 1973. He had been asked to join in 1967 but declined because he was too busy and "didn't want to be involved with the government during the Vietnam War." When he was asked again, he accepted because the war was ending. He was still a little nervous about what Jason actually did, and about whether he would "end up feeling like I was a tool of the government," and so in his mind he gave Jason three years. After the first summer—which he spent "in generalized bewilderment about what Jason did"—he understood that Jason chose what it would work on. "So you can be an inadvertent tool of the government—they can use you in ways that you have no control over," he said. "But you can't be an ardent tool of the government."

Eventually, Abarbanel said, he became "convinced that Jason was very valuable and especially so quietly. I have the feeling that the same tension that Robert Oppenheimer expressed about creating nuclear weapons was there in people like Murph but not verbally expressed. And I think when the opportunity came to participate in tamping down the monster that had been released, they were glad to do it—this is all my opinion—and they wanted to do something to make the world that they found to be pretty scary, the postwar world, a better place. And I think they passed that on to my generation."

CHAPTER SIX

Changes

"There was certainly a sense that if Jason was going to work on important problems, that DARPA didn't own all of them. And so the question becomes, can other agencies bring problems to Jason which are more demanding or more interesting? I mean, it's easy to complain about DARPA directors. It's harder to get off your keister and go do something."

—Edward Frieman, interview, 2003

After the war in Vietnam, nobody liked scientists anymore. The largely antiwar public saw them as the military's partners in pursuing unjust and immoral wars. The growing environmental movement saw them as the people who helped industry pollute the air, poison the water, and turn the land brown. A sample of headlines in the early 1970s *New York Times* included: "25 Form Council to Make Scientists Aware of Consequences of Their Work"; "A Hippocratic Oath By Scientists Urged"; "Scientists Urged to Improve Image."

During the Johnson administration scientists started having trouble finding jobs; money going to research grew at a rate below that of inflation; money going to pure research now went to applied. Science that had been seen as meritocratic was now seen as elitist, and democratizing ran rampant. Money that had gone to the handful of top research universities got spread out to the nonelites. Appointments to PSAC became so democratic that only one of its members came from Cambridge; by 1968 President Johnson wouldn't even let PSAC eat in the White House dining hall. During the Nixon administration the trend intensified. Congress criticized the National Science Foundation for awarding fellowships to the top graduate students. The National Medal of Science was no longer awarded; two of the three president's science advisers resigned because they felt useless. By the end of Nixon's first term in 1972, he declined to reappoint PSAC. ARPA's budget, for a while, looked like it had fallen off a cliff.

Jason's "esprit de corps" was also in decline. "In the earlier years, there was a spirit that we could do wonders," said Sam Treiman, "and we discovered we couldn't always do wonders." And Jason's studies— as usual, concentrating on test ban treaties and missile defense— were fewer in number; the ARPA historians wrote that "Jason productivity" decreased significantly. ARPA wondered whether it was getting its money's worth and began asking for lists of Jason's studies and membership, and when the Jasons first joined and what they were paid. ARPA's new director, Stephen Lukasik, thought Jason was due for a change. "I was very concerned that the Jasons were frozen in time," he said, "that they were an excellent group circa 1960, we're now talking about ten years later and it's the same group. Meanwhile the world has changed."

Lukasik was more or less right. Since Jason's formation in 1960, it had added only a handful of new members; the average age of the membership, every year, got older by one year. Jason was slipping into the pattern of aging old-boy advisers that it had been created to circumvent. And the world in which Jason was operating was indeed changing drastically. In particular, the problems ARPA was studying had changed. ARPA was now funding a reasonable fraction of the advances in information technology; the Internet was born in the late 1960s as ARPAnet. Jason's original charter, on the other hand, included the phrase "minimum expenditures will be made for computers." Jason itself had no computer scientists, in fact, had no scientists interested in computers. Theoretical physicists at the time, and not only in Jason but generally, thought problems should be solved by thinking and not by messing around with computers. If ARPA wanted help with the new world of computers, Jason couldn't offer it.

ARPA was also increasingly interested in the technology of producing materials. Materials could now be designed for strength or lightness or heat resistance or flexibility or imperviousness or any other quality or combination of qualities that the military might want in its equipment, not to mention the materials that would allow smaller, faster circuits for computers. Jason had no materials scientists either. The head of ARPA's materials office told Lukasik that the Jasons "are not my kind of people because they don't know anything about materials. I don't need any quantum electrodynamicists, I need someone who knows about semiconductors."

Meanwhile within ARPA itself and in the science advising business nationally, the level of scientific expertise had risen. Ten years after ARPA's birth, it had its own staff of competent scientists who knew what ARPA's needs were and where to get help. And high-quality technical help was increasingly available throughout both government and industry. By now, federal agencies had their own internal cadres of scientists, as had the intelligence community. So had industries like Martin Marietta, Northrop, Lockheed, and Boeing. So had research centers with federal contracts, the FCRCs, like RAND and IDA; so had the university-associated laboratories with large defense contracts, like Lincoln Laboratory (associated with MIT) and the Jet Propulsion Laboratory (associated with Caltech). ARPA knew how to use this widespread expertise: for a problem whose solution required, say, propulsion, chemistry, artificial intelligence, and computer architecture, ARPA could call on the relevant scientists in government, industry, and academia. By 1970, Lukasik said, "ARPA could put together a dream team on any problem." By comparison, Jason sinks, he said, "not because Jason sinks, it's just that the sea is rising."

Stephen J. Lukasik, like the early Jasons, was trained as a physicist. Coming out of World War II, "physicists were special, magic, wonderful," he said. "The reason why I'm a physicist is that I was also influenced by the World War II nuclear weapons story." Lukasik got his Ph.D. from MIT in physics in the mid-1950s, about the same time that the early Jasons were settling into tenured jobs at Princeton, Stanford, and Berkeley. Though he had begun with the private Nobel hopes of most young physicists, he said, "it became clear that I was competent and had a number of virtues, but probably a Nobel laureate in physics was not likely to be one of them." Moreover, he said, he was not enchanted with the academic goals of tenure, research, and publication: "I never quite saw that," he said. What he saw instead was a "professional ethic of simply understanding," he said, and "doing a good job and helping where I could."

After he graduated from MIT, he worked at the Stevens Institute of Technology studying fluid dynamics. Then in 1966 he took a job at ARPA. Not only was ARPA a "new vibrant agency," he said, but now he could participate "not as a spender of money but as a decider of what it should be spent on." Lukasik became head of ARPA's program of detecting nuclear tests, called Vela; then he became deputy director of ARPA, then the director. His philosophy of management, wrote the ARPA historians, "emphasized the authority of the director."

When Lukasik arrived at ARPA, he already knew who the Jasons were. "They were in the intellectual forefront of everything we were trying to do to prevent technological surprise," he said. "Technological surprise" was the current government jargon for what had happened when Sputnik went up. Jasons "all had egos so they all knew how good each of them was," he said. "If a less-capable physicist had been injected into the Jasons, they would have rejected him—'oh you know, Charlie's an idiot, let him bring the coffee.'" Lukasik's mental model of Jason was a fraternity that would, through its connections and its own judgment, decide what ARPA needed to do. "Do an experiment," he said. "Give that model to sociologists and ask them what would have happened in 1966 when Lukasik questioned it." In other words, put together a smart, independent group with a smart, authoritative manager and see what happens. You wouldn't have to ask the sociologists.

Lukasik began questioning Jason the first time he met them, in 1967, when he was still head of the Vela nuclear testing program. Vela was the second, besides the antiballistic missile program called Defender, of ARPA's two main missions. By now, the nuclear countries were able to detect illegal explosions in space, in the atmosphere, and under water; and they had made enough progress on seismic detection that they could begin negotiating for a test ban treaty that would include tests underground.

Hal Lewis had just become Jason's chair. In early 1967 Lewis walked into Lukasik's office, Lukasik said, "and told me that he and the Jasons were going to help me that summer and they were going to do X and Y and Z, and in order to do that they wanted access to places A, B, and C." What the Jasons wanted to work on was the part of the treaty dealing with on-site inspection. That is, could one country, detecting seismic waves that might have come from underground tests in another country, then travel to the suspicious site and see for themselves? Could they bring in instruments? What could the instruments detect about underground explosions?

Meanwhile Lukasik had just learned about the newly developed, highly classified reconnaissance satellites, spy satellites. The satellites were so secret, their classification was separated into compartments: one compartment was their very existence; another compartment was their results; a third compartment was the details of their operation. Lukasik had just been cleared into two of the three compartments. So he knew that satellite reconnaissance of suspect locations meant that on-site inspections were unnecessary. In fact, when Hal Lewis walked into Lukasik's office saying that Jason would do X, Y, and Z about an on-site inspection program, Lukasik was about to cancel the program.

Not only did Lukasik not need the Jasons' project, he didn't trust their motives. Jasons were "obviously pro–test ban," he said, and he thought they were worrying that "the on-site inspection thing is slipping off the table by reasons of impossibility and they're trying to save it." So Lukasik reacted to Lewis, he said, "by getting my back up. And I told him, 'Thank you very much, but I don't want your help.' When someone walks into your office and says, 'I'm going to do X for you,' the implication is, you're not smart enough. It was socially inept, it was a put-down, it was arrogant. I remember it was a meeting that went on for like forty minutes, at which point I remember Lewis closing the thing off by saying, 'Well, I'll have to discuss this further.' And I said, 'Fine, go discuss it further.' "

A week or so later Lewis was back in Lukasik's office, this time with another Jason, Gordon MacDonald. MacDonald had not only been a full-time academic, he was also vice president for research at IDA, Jason's host organization and—with ARPA—its other parent. So he knew how to handle the cultural, personal, and professional tensions between government and academic scientists. In the process the parties figured out that all three of them were cleared to know about satellite reconnaissance. "It was a marvelous example of secrecy and how it absolutely complicates communications," Lukasik said. The second meeting ended less acrimoniously, and they all agreed that Jasons should not study satellite reconnaissance because Jasons worked best in groups and the compartmented clearances were too scarce. "So that was my first introduction to the Jasons," Lukasik said.

The Jasons say that Lukasik's initial reaction to them is just the first step in a cycle of reactions common to all ARPA directors. ARPA directors tend to stay in office for two or three years, and their attitudes toward Jason all tend to follow the same trajectory. "Directors of ARPA often start out hostile," said Will Happer, "and then they realize after a year or two that 'you know, the only honest advice I'm getting is from Jason.' And eventually they become warm supporters of Jason." Or as Ed Frieman said, they'd start out asking, " 'Why are we giving those turkeys two million dollars,' and then after a while they would find out we were pussycats." Jasons mention this turkeys-to-pussycats story so often that it amounts to a family myth.

ARPA directors say that once they get to know Jason, they understand how to use it. Charles Herzfeld was ARPA's director from 1965 to 1967: "I had to explain why I was spending—it was then about $440,000 a year—on something that, as far as anyone could see, had no clear-cut purpose." After a year or so, however, Herzfeld came to understand that Jason "produced a major change in defense technology every year or so." And so, he said, he "went on the offensive" and told Jason's critics within ARPA, " 'Do you know anywhere else where I could buy a major revolution in defense technology every year for $450,000?' And that began to shut up some of the critics."

The tension between ARPA and Jason, between government and academic scientists, between money spent and value given, between insider and outsider, is implicit but obvious. The academic outsiders could see the government's problems rationally and solve them brilliantly but were impatient with ignorance and intellectual muddiness, and they often didn't know about the political trade-offs that got things done. The government insiders had the responsibility for the problems and saw the academics as sources of inventive solutions but thought them naïve, personally arrogant, and oversanguine about the extent of their influence. Herzfeld said he had to mediate between Jason and other governmental decision-makers: "You have to warn them and say, 'Look, you are going to hear some things that will sound crazy, but please listen, because they are not crazy. They may be wrong, but they're not crazy,'" he said. "And conversely, one has to sensitize the Jasons when one brings someone in, that please don't think the person is an idiot because he cannot integrate a simple differential equation." The tension between powerful insiders and independent outsiders has a solution that's obvious—Herzfeld's mutual understanding and forbearance—but never permanent. The tension between Jason and ARPA was heightened because Jason had no other sponsors; ARPA was the only game in town.

By 1971, when Lukasik became ARPA's director and decided Jason needed renewal, he specifically wanted Jason to evaluate its members' participation, rotate its membership, rotate its chairmen (in eleven years, Jason had had only two, Murph and Hal Lewis), and branch out beyond theoretical physicists. "Rather than say, 'Fire the bunch of them and get the right bunch of people,'" Lukasik said, "I was really quite gentle and moderate. I said, 'Give everyone a fixed term, when it's over review how much they've done, review the balance of skills in Jason, and decide if maybe you ought to rotate someone off.' Those are all good things." Lukasik couldn't mandate his desires, though, because ARPA's arrangement with Jason was to fund it, not to manage it. "Strictly speaking," Lukasik said, "ARPA gave the money to IDA and then IDA ran the Jasons." So Lukasik went to IDA and said in essence that Dad, who was the breadwinner, thought Mom should get the boy to shape up.

Mom said no. IDA's president, Alexander Flax, didn't want to interfere: Jasons, he said, "were intended to be as independent as possible." Independence is regarded as useful to the government, he said, though the regard is grudging: "I mean, people don't like to be criticized or even get a lukewarm endorsement of their favorite program. They want enthusiasm. But they don't always get that from the Jasons." That said, Flax agreed with Lukasik that Jason should "refresh its membership and broaden its outlook" but that such suggestions should be made gently. Jasons "were an anomaly, and they were set up to be an anomaly," he said, and "it would not be constructive to force this on the Jasons." What Lukasik understood from Flax was that IDA's approach to the Jasons was to arrange clearances, publish reports, set up meetings, supply lunches, "see that they don't get into trouble, pick up after the elephants, do your best," he said. "I think that IDA was not going to push Jason."

The obvious solution for Lukasik was to find Jason a pushier mother. So he did what he called "a clever bureaucratic thing, i.e., fix a management problem while also easing other management problems." One of Lukasik's management problems when he was ARPA's deputy director had been "the ceiling problem." Congress set a ceiling on the total amount that the Department of Defense could spend on its contracts with these FCRCs, and DoD would then apportion that amount out to the services, who would each in turn spend some at IDA, some at RAND, and so on. ARPA could allot IDA, say, $5 million, and IDA in turn would spend maybe $500,000 of it on Jason. So Lukasik as director remembered what he'd learned as deputy and thought, "If we moved the Jasons out, IDA could keep the ceiling. We could take the Jasons and attach them to almost any other perfectly respectable non-FCRC organization"—that is, one that had no ceiling. The bureaucratic cleverness was not so much that IDA would now have freed-up ceiling money from ARPA, as that Jason could be reformed: to change an institution, Lukasik said, "the bureaucratic principle is, you move it and change it in transit." In other words, he said, Jason might get on the train in business clothes, but it would arrive at the resort in shorts.

So in 1973 Jason left IDA and was installed at the Stanford

Research Institute, called SRI, a research center that didn't happen to be classified as an FCRC and so was immune from the ceiling requirement. At the time SRI was in the process of cutting its historical ties with Stanford University; it specialized in information technology and had the honor of being one of the first four nodes of ARPAnet. Its headquarters were in Menlo Park, California, but it had a satellite office in Washington, D.C.

The move cost ARPA (which had meanwhile renamed itself DARPA, Defense Advanced Research Projects Agency) over $500,000: it left IDA's ceiling where it was and so had to fund SRI to take on Jason. The outcome, as far as Lukasik was concerned, was a success. "The Jasons became absolutely useful and well managed," he said. "By the time I left in 1974, as far as I was concerned, that was a good group of people. Smart, effective, working on good things, well coupled-in." That year Jason did a study for Lukasik on a special-purpose computer to simulate the transition from smooth to turbulent flow of fluid around, say, a submarine. Lukasik had wanted to know whether the state-of-the-art computers could handle such large simulations; Jason said they couldn't— "Jasons put the nails into the coffin of that idea," he said—and Lukasik dropped the project. So does Lukasik fit Jason's story about every new director of (D)ARPA coming in a doubter and leaving a believer? "Let me say," he said, "it certainly applies to me."

In general the Jasons seemed to think the change was simply good sense. Because the first group was the same generation, stayed the same size, and maintained the same expertise, Jason at some point was bound to be a little elderly, out of touch, and in need of renewal. So Jason formally became a little more professional. A chairman, elected by the steering committee, would serve for three years and could have no more than two terms. Members would neither have fixed terms nor be formally rotated. But what had been an informal trial period for new members became, in Jason's

1976 charter, an initial two-year term. And a membership committee would now periodically review the members for performance
and for "distribution of expertise." Members could, if they were
less active, ask or be asked to become senior advisers and be allotted maybe fifteen days of work and be invited to the spring and fall
meetings and the summer wrap-up. Between 1966 and 1973 Jason
had added nine new members, seven of whom were physicists, one
an oceanographer, one a biologist. In 1974 alone it added six new
members: four physicists, a computer scientist, and an electrical
engineer. The changes seem sensible and even obvious.

Once Jason began branching out into new disciplines—and
maybe to avoid being captive to DARPA—it also branched out
into new sponsors. Frieman said that "there was a sense that the
whole world didn't fit into the DARPA thimble." In 1977 Jason's
summer studies—though the wide majority were for DARPA—
were for the CIA, the Defense Nuclear Agency, NASA, the navy, the
National Science Foundation, and ERDA, the Energy Research and
Development Administration, which was about to turn into the
Department of Energy.

Jason's changes sound bureaucratic and unremarkable—a
few nonphysicists, some new sponsors—but tailoring itself to its
customers turned out to have surprisingly large consequences. The
group became less cohesive and its studies more diverse. More of
its studies required compartmented classifications, and those Jasons in the compartments could not talk to those outside them.
And some of their studies now had nothing whatever to do with
defense or the military and were in no way classified: "it was obvious that we could do more interesting things than fight the Cold
War," said Freeman Dyson.

The biggest new sponsor was the navy. Jason had already been working on projects for the navy (and for the CIA), but those projects

were run through and sponsored by DARPA—Nick Christofilos's Bassoon/Sanguine projects, for example, were for the navy via DARPA—and were relatively few. Now that Jason had a separate contract with the navy, the studies increased. The connection began, Frieman said, with "a set of very troublesome issues vis-à-vis the Soviet submarine force and what threat it posed to the United States. And lots and tons of all sorts of crazy things that were picked up in intelligence."

For Jasons, working for the navy meant they had to learn oceanography. And learning something new is what Jasons, physicists, and scientists in general evermore love. "The ocean is largely stratified," said Will Happer, now faculty at Columbia and a new Jason. "It's not like the air. Today the air is churning around because the sun is heating the ground, and up to ten kilometers, it mixes very quickly. Ocean doesn't do that. Of course, as a physicist, nobody ever told me about those facts. So the lesson in that physics was worth the whole summer."

Educating Jasons in their new subject was the job of Walter Munk, an eminent oceanographer at the Scripps Institution of Oceanography and one of the rare early nonphysicist Jasons: "I am an oddity at Jason," Munk said. He enjoyed the Jasons' enchantment with the ocean and thought they were "wonderful to work with" but "naïve about the way in which you work on observational and natural sciences." Henry Abarbanel's first Jason meeting was at Munk's house: "Walter trained a cadre of younger people, including myself. I didn't know from the ocean—I mean, I know it's out there."

The navy's problems are interrelated: how to detect enemy submarines, how to hide ours, how to communicate with ours when hidden. The principal methods of addressing all these problems involved sound. Can you detect enemy submarines by listening for their engine noise? Can you quiet your own submarines? If you sunk arrays of listening devices—the sonobuoys that were the basis for some of the electronic barrier's sensors—into the ocean, how would you tell a submarine from any other noise-maker? Can you find submarines by broadcasting sound into the ocean and listening for certain echoes? How far can sound travel in the ocean? In fact, how does sound travel in the ocean at all? Does sound travel differently with changes in the ocean?

One such change, as Happer said, is that the ocean isn't mixed together but is stratified, has layers of differing densities, lighter water sitting on top of heavier water. At the interface between the heavier and lighter layers, said Curt Callan, a Princeton physicist who'd joined Jason in 1967, "you have waves that travel very very slowly—those are internal waves." Internal waves dramatically affect the way sound travels through the ocean: "the farther sound goes," Callan said, "the more distorted it becomes." Not incidentally, a traveling submarine also creates internal waves. Callan, Munk, and other Jasons wrote a report called "Generation and Airborne Detection of Internal Waves from an Object Moving Through a Stratified Ocean."

Different parts of the ocean also have differing temperatures, and temperature also affects the way sound travels: faster in warm water, slower in cold. In the 1977 summer study Munk shared an office with a new Jason, a colleague named Carl Wunsch from MIT. Wunsch had been working on a kind of mathematics, called inverse methods, that would make sense out of huge amounts of data. Sitting in their Jason office, Munk and Wunsch discovered that if they put together what Wunsch knew about inverse methods with what Munk knew about sound indicating temperature, they could make a three-dimensional map of the ocean's temperatures. "Dick Garwin wandered in to see what we were up to," Wunsch said. "And when we described to him what we were thinking about doing, Dick said, 'Oh, you've just reinvented tomography.'" Tomography is the T in CAT scan, a technology that stacks up successive two-dimensional X-rays in a computer and ends up with a three-dimensional image.

In the case of acoustic tomography, the X-rays are sound waves broadcast through the ocean, and the resulting image is of the sound's different speeds, and therefore a three-dimensional map of the ocean's different temperatures.

"Now, anything to do with acoustics in the ocean has military implications," Wunsch said. But the resulting Jason report, "Preliminary Report on Ocean Acoustic Monitoring," also opened up a new scientific field: "I think it's the first time the words appear, ocean acoustic tomography," Wunsch said. Ocean acoustic tomography, first done in the late 1970s, has since become big science. It continues through projects like the government-funded, multi-institutional $35 million Acoustic Thermometry of Ocean Climate project. Done with loudspeakers and arrays of listening devices, it charts how the temperatures in the ocean are distributed and how, over time, they change. Most important, since the oceans soak up most of the heat in the planet's energy budget, ocean acoustic tomography also charts the planet's warming.

But the navy regularly considered submarine-detection methods other than sound, like radar or radioactivity, called nonacoustic antisubmarine warfare; Munk said Jason occasionally called it "unsound methods of submarine warfare." One famous nonacoustic method the navy proposed was to use the subatomic particles called neutrinos that submarines' nuclear-powered engines gave off. This method was suggested often enough that Jasons finally wrote a report called "Neutrino Detection Primer" that outlined in great mathematical detail and with some exasperation the widely known fact that because neutrinos are diverted by neither the electromagnetic force nor the gravitational force, both of which govern life on earth, they travel unswervingly through the whole planet. Capturing them is correspondingly improbable, and detecting those coming off submarines is nearly impossible. The proposal, the Jasons write, is "an example of a phenomenon which occurs frequently in the history of science: a clever and beautiful idea killed by stubborn facts."

The navy is unusually tight about its classification rules and unusually stingy about its clearances, which, like the clearances for spy satellites, tended to be compartmented with only a few people allowed into each compartment. So although all Jasons had top secret clearances, only a few were cleared into the navy's various compartments. As a result, those few tended to work mostly on navy studies; they were called the Jason Navy.

The Jason Navy had around ten members, including Ed Frieman, Bill Nierenberg, Ken Watson, Ken Case, Curt Callan, and Walter Munk. Though nonnavy Jasons worked on navy studies, and Navy Jasons worked on nonnavy studies, the group wasn't particularly fluid. Over the years, the relationship between Jason and the navy grew close enough that the Jason Navy almost became the navy's corporate memory. Few of the people from any of the armed services who receive Jason's reports stay in their positions for more than two or three years; the Jason Navy, though it added younger members, was stable. And as a result, said Case, "let's suppose they've observed some Russian ships doing something. They would come to us: 'what could they be doing?' And we go through our memories, the kinds of things the Russians have been working on. So we have a memory."

Because of the Jason Navy, for the first time in Jason's routine practice, not all Jasons could talk to all other Jasons. Since Jason's birth, its most striking characteristic, besides its intelligence, was its collegiality. Scientists came into Jason with more or less the same education, language, manners, and habits of work. They knew one another professionally; many were collaborators; they had been one another's students or teachers; their jobs were at the same few universities. As Jasons they tended to work in small groups, but people could join or leave, or join for a while then leave; and each group knew what all the other groups were working on. They could go to all the briefings; they could walk down the hall and ask for help. The way Jason operated reminds you that *colleague, college,* and *collective* are fundamentally the same word. Jason went past personalities and past individuals to that entity, the community: the whole that is greater than the sum of its parts.

With compartmentalized classification and the Jason Navy, the community cracked a little. Compartmentalization, said Sid Drell, meant that "now we know less and less about what other people are doing." And that was "a good example of a harmful trend," he said. "I believe firmly that if we have a more open exchange among good scientists, good ideas get ventilated, criticized, and created. So I think it's all harmful. But it's a growing trend, an insidious trend, perfidious trend." Ed Frieman, who was in the Jason Navy from the beginning, said, "It's led to many hours of discussion at steering committee meetings. It became very divisive to Jason."

Though Jason's other new sponsors weren't as picky about clearances as the navy, the Jason community nevertheless began gradually forming patterns of collaboration that could be called cliques. Certain Jasons had the background or the taste for the studies proposed by certain sponsors, and therefore tended to cluster. And as Jason studies expanded out into different fields, the members of those clusters began to accumulate specialized educations, and the clusters, in turn, began to tighten. These clusters were not threatening—if a clique started to get exclusive, said one Jason, the steering committee "executed some quiet diplomacy"—but they were noticeable enough to be given informal names: the Jason Navy, the arms controllers, the boy spies, and a new one, the climate group. Jason could be thought of as expanding out into the world. Jason also could be thought of as undergoing condensation, like mist into water drops or gas clouds into stars.

The new climate group was large and, because no clearances were necessary, varying. Henry Abarbanel and Gordon MacDonald were more or less the group's constant core; regular members included William Nierenberg, Ed Frieman, and Freeman Dyson. The first climate study, in 1977, had no sponsor. Frieman, who was now Jason's chair, said that since Jason's budget wasn't necessarily tied to specific projects, the first climate study could be done with some "free money." Jason decided to study the subject on its own.

Though Jasons liked climate because it was new and interesting, to the war-averse, post-Vietnam Jasons, climate was also innocent. Frieman said that after Vietnam, Jason had "a sense that maybe we should move out of the defense-dominated domain. We used to have meetings that were called Whither Jason—*whither* with an *h*, not an *i*—and we would say, 'We're designing all this hotshot warfare stuff. Maybe we should do something for the rest of the world.' And so climate came up in that context." Gordon MacDonald was even more explicit: "Jasons wanted to work on things so that they could—and I'll put it a little too bluntly—go back and tell their colleagues that they were really working on peaceful rather than war-related projects."

Moreover, to the public and in the media, climate was an increasingly hot topic. The carbon dioxide we were adding to the atmosphere was warming the planet. The sulfur from burning fossil fuel made the rain acid. By the end of the 1970s everyone knew the phrases: global warming, acid rain, greenhouse effect. But scientists still weren't sure whether the climate was changing or not, and if so, in what direction. "We were at the beginning of the global warming era," Frieman said, "and had very little idea of where this was all going to go."

The climate Jasons had no resident expert, as the Jason Navy had in Walter Munk, so in the spring of 1977 they convened a two-day meeting in Boulder at the National Center for Atmospheric Research and invited government and academic experts to talk about what changes humans were causing in the atmosphere. Levels of carbon dioxide in the atmosphere were steadily increasing, making the atmosphere increasingly likely to trap heat, and the experts thought that global temperatures would rise accordingly. One way to understand the whole system, the experts said, was to model it in a computer.

Jason decided to make its own climate model. It was called "Features of Energy-Budget Climate Models: An Example of Weather-Driven Climate Stability," but Frieman referred to it, with a certain tone in his voice, as the Jason Model of the World. "It was a model of global warming," Frieman said, "a very simple model." Climate scientists in both academia and the government laboratories had been running computer models of global warming for years, but in spite of growing environmental awareness and well-funded climate studies, Frieman said, "the models were pretty crummy." If fifteen different models were asked the same question, they'd give fifteen different answers. To make matters worse, the models were so complex that when they disagreed, Frieman said, "it was difficult to sort out what was wrong. Did you have the wrong mathematical formulation, or was the physics wrong, or the chemistry was wrong?" The Jasons thought that if they simplified the model, they might learn which of all its parts was the most critical.

Simplifying the model is a physics trick. Physicists prefer to study a system from the ground up—that is, by knowing how the parts behave, they can understand the whole: neutrons and protons behave this way, so an atomic nucleus must behave that way. But some systems—a magnet, a turbulent fluid—are too complicated to be understood from the ground up. In those cases, physicists find ways to average the behavior of the parts: in the case of climate, an average atmosphere, an average land, an average ocean. Change one of these averaged parts—make the atmosphere more likely to trap heat—and see how the whole climate responds. Physicists call this coarse-graining; they're famous for it, and they're proud of their ability to do it. They also call it back-of-the-envelope or desert-island physics; it's a variant of what Fermi did with his pieces of paper to get the yield of the Trinity explosion. Coarse-graining works if the average atmosphere you've picked is reasonably representative of the actual atmosphere—that is, Murph said, only if you "figure out what can be left out without losing the essence of what you're trying to describe." In short, it works only if you know how coarse the grains can be.

The Jason Model of the World changed the atmosphere by adding more carbon dioxide, and as it happened, that worked. "You put carbon dioxide in, and the temperature went not through the floor but through the roof," Frieman said. "I mean, it worked in the sense that it looked like a reasonable view of what was going on in the world." Abarbanel led the study: "We did a little model-making and we had fun, a lot of fun. It was a nice little physics project." The study's "one hiccup," Frieman said, was that the community of climate scientists got upset: they said, "What the hell are you guys doing? You're not climate scientists. This is our business"—meaning, not "climate is our territory," but "climate isn't that simple," the grains are too coarse. Fine-grainers always say this about coarse-grainers, and sometimes they're not wrong.

In 1979 Frieman left the Princeton Plasma Physics Laboratory, the fusion research lab that had succeeded Project Matterhorn; took a leave from Jason; and became the head of the Office of Energy Research at the newly created Department of Energy. For the next two years he commissioned Jason to do a series of studies on the relationship between carbon dioxide and global warming. After Frieman left office, the Department of Energy continued to ask Jason for climate studies, this time on the less complex problem of acid rain. The smoke from burning fossil fuel contained sulfur dioxide; when sulfur dioxide got into the atmosphere, it combined with water to make sulfuric acid, which then came down with the rain. Lakes were poisoned, fish died, forests turned gray, and statues and buildings dissolved.

One of Jason's acid rain studies came to the attention of a mechanical engineer at Brookhaven National Laboratory named Aristides Patrinos, called Ari. In the early 1980s Patrinos was working on models of acid rain when his boss asked him to review a Jason report on the subject, and Patrinos said, "Who's Jason?" Patrinos read the report—he doesn't remember its name—and thought, like the fine-grainer he was, that it was "extremely simple-minded." He also thought it was "pretty refreshing." The chemical reactions taking place in the atmosphere were not as straightforward as Jason thought, but he liked their noninsider approach. Brookhaven's final review of the Jason report was dismissive: Patrinos suspected his colleagues thought that because Jasons weren't climate scientists, they weren't worth hearing out.

The climate scientists' complaints raised new questions for Jason. The academic climate community had been working on the same questions Jason studied long before Jason got into the field; and ever since then the community had gotten only larger and better funded. What could Jason bring to the field that the climate scientists could not?

Carl Wunsch—who had worked on navy studies with Walter Munk but who, as an oceanographer, also knew about climate and Jason's studies—didn't think Jason brought anything to climate at all. "I developed a distaste for some of what was going on in Jason," he said, and in 1981, after three years in Jason, he resigned. The first thing that bothered Wunsch was that scientists were already paid by taxpayers to do the studies that Jason was also getting paid to do. "I didn't think Jason should be paying me as a consultant to work on something that I would happily take money from the National Science Foundation or the Office of Naval Research to do, which I did," he said. "And I thought they were abusing Jason."

Wunsch also agreed with the climate scientists that the problem of global warming was too complicated to be solved in a series of six-week summer studies. "To me, it was a bunch of amateurs working in my field," he said. "They were getting on the phone to the people who really were the experts, and then they would put out this Jason report." Wunsch wrote a long letter about his worries to Richard Garwin, who was now chair. Garwin called Wunsch and said he understood but could do nothing about it.

Wunsch's last reason for resigning was personal and slightly more complimentary to Jason: on nonocean, nonclimate problems, "I couldn't compete with people like Murph, really quick thought-patterns, looks at a problem, goes bang. Well, I couldn't do that. I'd have to go home and think and try things and look in the library. So these are brilliant people."

Wunsch is to be taken seriously. He is as eminent an oceanographer as Munk, and his humility about his brilliance relative to Murph's notwithstanding, he in no way gives the impression of intellectual insecurity. He didn't dislike Jason; he said Jason has some "very dedicated, very smart people," and that he's "always been glad to know there were some smart people who were actually willing to tell the Defense Department things it did not want to hear." Clearly, though, he thought the overlap between his field and Jason's studies was scientifically redundant and a little murky. Besides, he said, "I did resent this 'the ocean is a wonderful summer playground for smart physicists who can do it in their spare time.' You may know this thing called physics arrogance. It's real. And you know, I just didn't need it. I didn't need it."

Jason eventually seemed to agree with Wunsch's arguments and the climate scientists' complaints, and for a number of years in the mid-1980s it did no climate studies. "After the initial flurry," said Frieman, "people realized it was a tough problem and Jason was unlikely to do anything fundamental. There's a big community out there."

Then in 1986 Patrinos moved from Brookhaven to DoE, working under a director who happened to be a physicist and who thought that climate problems had been "captured by a smaller scientific community and that the issue was too important to leave it to the specialists in atmospheric science." The director said the problem deserved both open-mindedness and rigor and suggested that Patrinos "look into Jason," Patrinos said. "And then it clicked. I remembered Jason. And at that time Gordon MacDonald happened to be in the area." MacDonald was now chief scientist at the Mitre Corporation—another FCRC like IDA—in the suburbs of Washington, D.C., "so he was local," Patrinos said. By now DoE was setting up a new interagency research program—the Atmospheric Radiation Measurement program, called, of course, ARM—to further study climate problems. Patrinos and MacDonald got together, and MacDonald recommended that Jason advise Patrinos specifically on ARM problems. "Gordon led the charge," Patrinos said: as a geologist, MacDonald had the background for it, and besides, he thought climate was as much an issue of national security as defense was.

The ARM program was set up specifically to address the climate-modeling problem: that after a decade of progress, the models were still huge and complex and still didn't work. One reason they didn't work was that no one had yet done the necessary long-term, large-scale, systematic measurements of the variables—clouds, heat, the atmosphere's aerosols, humidity, the greenhouse gases—that affect climate. Put inexact data into a computer model, and you will get inaccurate predictions. "It was a physics problem, it was a measurement problem," Patrinos said, "and it required instrumentation that had been beyond the capabilities at the time." ARM's aim was first to take the measurements and next to merge them all in a computer model and see whether the planet was warming or cooling.

Toward that end and with the guidance of Jason, ARM began taking measurements that had never been taken before. It set up several sites—in the southern Great Plains, in the tropical western Pacific, on Alaska's north slope—each representative of a certain range of climatic conditions. At each site measurements are made by instruments on the ground, on satellites, and in various kinds of manned and unmanned aircraft. "Because of Jason," Patrinos said, "we ended up using unmanned aerospace vehicles for some of the measurements at the top of the atmosphere."

Jasons not only helped set up the ARM program, they also reviewed it several times for its effectiveness, its science, and its direction for the future; on the whole, they liked it. "In effect," said Abarbanel, "we helped set up what is the only long-term, excellently instrumented geophysical observing site for atmospheric sciences dedicated to collecting data over decades. I'm really proud of having been part of that. Really proud." The ARM program, Patrinos said, "can honestly be called and considered a child of Jason."

Patrinos is pretty much an all-around Jason fan. Granted, he says, "sometimes they had been pretty sloppy with respect to the reports they've written. This is not a group that prides itself on having accounted for everything they've done." What Patrinos enjoyed most about working with Jasons was "the experience of being out there with these guys for a week, duking it out on these issues and then subsequently going out to dinner and continuing the discussion." Patrinos said the experience was like an academic symposium, "but extended over days and a lot more acrimonious and lively. I also confess, I was intrigued by this notion of a bunch of very smart guys getting together and thrashing out issues. It was just heaven for me."

Patrinos said that Jason brought a hard eye to what can be a soft subject. In climate studies, he said, "what you measure is so complex, you tend to give up on the rigor. Because any condition in the atmosphere can never be the same again, the experiments that you do can never be exactly duplicated. So because there was this uncertainty, people generally didn't take measurements to the degree necessary. I think Jason's biggest contribution was bringing in the rigor."

As it happened, Garwin had agreed with Wunsch that Jason should get out of the climate business, but for a different reason: he thought that Jason's time and the taxpayers' money would be better spent on classified problems "because fewer objective people were doing that." Studies for DARPA and the navy had always been split between classified and unclassified, and Jasons had always argued about which deserved the most concentration. But now with Department of Energy and the climate studies, some studies were wide open in all their aspects, and for a while, anyway, the balance seemed to shift toward the unclassified, thereby exacerbating the argument.

The relative split between classified and unclassified studies was, and still is, unclear. Jasons' estimates of the fraction of its studies that are classified ranges from half to two-thirds to three-quarters to "most." One reason for the range is that the fraction has varied with time—it was high during the Cold War and lower post–Cold War. Another reason is that counting the relative number of studies doesn't count man-hours: unclassified studies like climate tend to be wide-ranging, long, and take the time of many Jasons; the more classified the study, the fewer the Jasons who work on it. Another reason is that classified studies occasionally result in unclassified reports. Garwin thinks that the fraction of classified studies is less than three-quarters and that the fraction of classified reports is even less.

The exact split among the Jasons, who after all each decide for themselves between classified and unclassified studies, is also unclear. I think no Jasons are all white—that is, never work on classified studies. Most are varying shades of gray; the lighter-gray Jasons, said Richard Muller, a physicist from Berkeley who joined Jason in 1972 and who is himself darker gray, just "don't like the hassle." Some Jasons are nearly all black, but they don't necessarily identify themselves.

The argument for concentrating on unclassified studies is they're personally and professionally congenial. Jasons take classification seriously and obviously have no objections to working on classified subjects, but they're not uniformly delighted to do so. Jasons who work on climate don't have to be careful what they talk about, don't feel isolated from other Jasons, don't have to worry about its moral implications, and can share what they learn with their students and colleagues. No scientist likes knowing things that are unshareable. Science without shared knowledge has no rigor, no foundation, and no future. Having unshareable knowledge particularly hurts the relationship between scientists and their graduate students. "I mean, I'm their master, they're my apprentices," Drell said, "and this is a whole life experience. And now part of your life is cut off. And that is a burden."

The argument for sticking to classified studies is more pragmatic. Most unclassified science already has a large, active, and self-critical body of scientists working on it. Classified subjects, like antisubmarine warfare, are worked on by smaller, isolated communities of government scientists who cannot, like academic scientists, send their research out for review. So if these scientists want to know how their ideas sound to outsiders, Jason is nearly unique. Jasons already have top secret clearances and have been working on such problems for years. "What Jason has from having been around for forty years isn't particularily available to the government in other forms," said Paul Horowitz, a Jason at Harvard who tended toward the black. "I think, given a choice and finite resources and equally worthy problems, it probably makes more sense for Jason to lean toward the ones that someone else cannot do."

Moreover, Horowitz says—as do other Jasons—that "sometimes the more classified the project, the more stupid the idea because the less review it's gotten by anybody who knows anything. I remember one Jason colleague saying, 'We like to think that we don't save the government millions, we save the government billions.' I'm not vouching for that particular quantification, but I'm telling you that sometimes you get a really stupid one." Richard Muller said one reason for the stupid ones is that "if something is highly classified, there may only be a dozen people in the world who have thought about it," meaning that simple mistakes are easier to make and harder to catch. "It's like trying to write a book without having an editor," he said. "And I think that's where Jason is most valuable."

I've presented this as an argument, but it's really a discussion with at least two right answers. Jason was created to work on defense problems, which are most often classified. And many Jasons join and stay in Jason because they want to keep an eye on science's potentially harmful inventions, which in most cases are military and classified. And besides, everyone likes secrets and the Jasons say they're no exception. "There's a kind of glamour," Murph said, "of knowing secrets."

But no one thinks Jason should jettison unclassified studies. "The world is a-changing, you know," the blackish Horowitz said. "If the interesting problems the government has are at the unclassified level, and they need the help, and we've got the manpower, we should do it."

In 1981 Jason traded its home with SRI for another one with Mitre, the federal contractor that had originated at MIT, that charged lower overhead than SRI, and that was bigger and had an office handy to the sponsors in Washington, D.C. New members continued coming in, again at the usual rate of one to three per year. One was Jason's first woman member, Claire Max. In 1979 Sam Treiman, who had dropped out over the Vietnam studies in 1968, rejoined. Treiman noticed that in the interim his friends in Jason seemed to have become more a part of the establishment: "The older Jasons by now had had a lot of experience, a lot of contacts, knew their way around Washington better," he said. "On the whole, the age was more establishmentarian."

Meanwhile among these new, younger Jasons, mildly revolutionary thoughts had been ticking. Rich Muller had found out that the per diem fees were based on the members' university salaries, so that younger members from less munificent universities were paid less. Will Happer said that he'd never been sure how the fees were decided: "it all seemed very secretive and capricious." The reasoning was that older Jasons were more eminent and needed higher fees to counterbalance the higher consulting fees they could command. Muller argued with the steering committee that the difference in fees was "the source of some jealousy," and his argument "carried the day." After that all Jasons made $250 per day. Muller said the "real upside is that a source of disharmony was eliminated, and the system makes Jason more attractive for younger members."

In 1981 Jason set aside the afternoon of one of its summer study days for another Whither Jason meeting and discussed possible changes in its optimal size, its mix of members, the kinds of studies it should do, and its governance. One new Jason, William Press, a Harvard physicist who practiced a combination of astrophysics and computer science, thought that though some Jasons dropped out and others had joined, "Jason had not gotten the ethos that it should be self-renewing." Another Jason charted the Jasons by years of service and found the chart had two peaks, one at six years and a larger one at twenty-two. Jasons noticed that all six of their chairmen—Murph, Hal Lewis, Ken Watson, Ed Frieman, Richard Garwin, and now William Nierenberg—had been either primeval or early Jasons. "And we could see what would happen eventually," said Happer. "Our chairs would be seventy years old." Over the next few years Nierenberg, the chair, slowly began putting younger Jasons on the steering committee, either because he had personal feuds with most of the older members or because he liked the idea of self-renewal: probably both. During the summer study of 1984—Bill Press thought it was a kind of watershed—most projects were led by younger Jasons.

That same summer Nierenberg announced that Jason would hold a gala celebration of its twenty-fifth anniversary. Younger Jasons didn't much like the idea because it seemed self-congratulatory, but Nierenberg, who was known for being a persistent and uninterruptible talker, prevailed: "he was unshakeable in his resolve," Press said. In response, the younger Jasons—mostly Press and Muller egging each other on—decided to write a musical skit expressing their viewpoint and perform it at the anniversary banquet.

They swiped the plots from Lerner and Loewe's *My Fair Lady* and Gilbert and Sullivan's *HMS Pinafore* and merged them into *Jason's Golden Fleece*. Press, playing Richard Garwin, bet Curt Callan, playing Nierenberg, that he could pass off one of the younger Jasons as an air force general. The young Jason happened to be a curly-haired, T-shirted, multiopinioned astrophysicist, playing himself, whom Garwin trained to talk about requisitions, management techniques, and to say C-cubed-I, military shorthand for Command, Control, Communications, and Intelligence. The astrophysicist Jason, carrying his usual baguette instead of a general's baton, managed to fool the colonels and generals at a Jason cocktail party until the director of the CIA announced that his spy satellites had uncovered the imposture. Garwin was about to lose the bet, but then Mildred Goldberger entered and explained that years ago when she was running a child-care center, she'd exchanged the infant astrophysicist Jason with the infant General Abrahamson, now the director of Star Wars; as a result, Jason was now in charge of Star Wars. The play ended with Jasons singing, to the tune of "For He Is an Englishman," that Star Wars, "In spite of its temerity / And its blatant insincerity / It remains a Jason plan / It remains a JA-A-A-A-A-son plan." Muller wrote the dialogue and Press wrote most of the lyrics, which are clever and scan beautifully. Press happened to have complete recall of all Gilbert and Sullivan operas; he bought a rhyming dictionary, he said, and just got into the swing of it.

The first and only performance was at Jason's twenty-fifth anniversary banquet, in November 1984, after only one rehearsal. "The production values were strictly amateur," said Press. The acting was energetic and occasionally convincing; and the singing was often on key. Press, playing a nitpicky Garwin, sang a song called "Why Can't the Air Force Teach Their Briefers How to Think?" Callan, playing a self-important Nierenberg, couldn't get through a speech without talking about his old days in Turkey with not one but two passports. Happer played the director of the CIA, Paul Horowitz played a colonel, and Claire Max had no choice but to play Mildred. When a general asked if Mildred was a Jason, the Jason playing Hal Lewis said that Jason had no women, that "the genes for arrogance and physics are both on the Y chromosome." What outsiders might have seen as self-conscious silliness, insiders saw as testimony to their sense of community: the audience laughed at every joke. Jasons loved it.

Less than a year later, halfway through Nierenberg's second three-year term as chairman, the youthful steering committee held what Happer called "a little palace revolution" and decided Happer should be its "stalking horse," he said. "I was the tool to sort of break this monopoly of the original founders." The steering committee argued that a chairman-elect be appointed, Nierenberg finally agreed, and in January 1986 Happer became the first chairman of Jason's second generation. It wasn't so much a palace revolution, said Press, as "a generational change." Muller called it "the movement from the old guard to the young Jasons" and said it was "amazingly smooth." The old guard, said Happer, "didn't mind too much." He thought some of them might have been irritated but didn't think "there was any bitterness about it."

Jason, which had long since given up meeting on alternate coasts, now settled more or less permanently into La Jolla. It was convenient to the San Diego airport, the families liked the beaches, and for the old Jasons, it was a nice place to retire. At first they met at a girls' school called the Bishop's School, using the dorm rooms as offices. "It was a nice place," said Happer, "but you know, we would have briefers coming in from intelligence agencies who were used to very high security and they would see this girls' school. We had the buildings swept carefully before we started every summer, looking for electronic bugs. And we had a good staff of guards there twenty-four hours a day. But it was clear that this couldn't work indefinitely." Besides, Jasons said, the Bishop's School found it could make more money running tennis camps. So in 1986 Jason moved across town to General Atomics, which had a facility called a tank or a SCIF, for Sensitive Compartmented Information Facility, which is a windowless, locked, guarded, electromagnetically shielded room for discussing top secret work. "So it looks a lot less like a bunch of amateurs getting together for the summer than it used to," Happer said.

CHAPTER SEVEN

Matching

"I mean, every once in a while I get morally indignant about something the government is doing that doesn't make technical sense, and I suggest to the rest of my colleagues at Jason that we do a summer study. And they tactfully point out that there's no point in doing it if nobody's going to listen."

—Claire Max, interview, April 26, 2003

With its injection of youth and diversity, Jasons settled down to work, and by the mid-1980s the number of their studies had increased dramatically. Despite their changes, much of what they worked on was still related to the Cold War and missile defense. Of all of the clusters, the missile-defense Jasons were the oldest, and probably over time, had included most Jasons; it was really hardly a cluster at all.

Back in 1972 the United States and the Soviet Union had signed the Anti-Ballistic Missile, or ABM, treaty to limit the number of missiles—two hundred—that each side could have and the number of places—two—the system could protect; two years later the countries cut those numbers in half. The idea of the treaty was to prevent one side from developing a missile defense so complete that it could wipe out the less-defended side and still feel protected against reprisals. While reducing the number of missiles was unquestionably a step in the right direction, the treaty didn't slow the

Cold War. Only the number of missiles was limited, not the number of warheads, and both sides were busy building missiles that could each carry several independently targetable warheads. The United States thought that in numbers of warheads, the Soviet Union had a considerable lead, and by the late 1970s it was worrying about a "window of vulnerability," meaning that the Soviet Union might think it could get away with a preemptive first strike.

The focus of the U.S. worry was the air force's new missile, called MX for missile experimental, also called the Peacekeeper. The MXs, each of which had ten warheads, were to replace the Minuteman missiles, which had only one to three warheads each and were stored in underground silos. The Minutemen were deemed vulnerable to being wiped out if the Soviet Union were to strike first with its own multiple-warheaded missiles at those underground silos. At first the air force wanted to put the new MXs into the old Minuteman silos, thereby—as Richard Garwin pointed out—replacing Minuteman vulnerability with MX vulnerability.

So the air force began a hunt for places to base the MXs where they'd be invulnerable. It came up with around forty different proposals for what it called MX basing systems, most of which kept the missiles on the move so they'd be hard to target. The basing systems, Freeman Dyson wrote, included "tunnels, trenches, deep underground burial; dispersal of mobile missiles on roads, on train tracks, and in airplanes; the multiple protective shelter or race track system, which threatened to cover large areas of Nevada and Utah with concrete, and the dense pack system, which would squeeze all the MX missiles together into a small area in Wyoming." All basing systems proposed seemed wildly expensive and equally unrealistic. Sid Drell called the dense pack system "dunce pack." The press and Congress called the MX basing systems a shell game. Jason did several studies to assess the MX basing systems and was similarly unimpressed. William Perry, the new DDR&E, the director of defense research and engineering, said missile basing was a hell of a problem.

Around 1979 Garwin and Drell got money from Perry's office for Jason to study its own MX basing system. "We had, I think, a very good study—I'm really proud of it," Garwin said, "that the only way to base the MX missile, and a very satisfactory way, was to put it on submarines." The navy already had its own missiles, called Tridents, on nuclear-powered submarines. The Trident's submarines were the size of buildings, partly because the nuclear reactor driving them was large, and partly because the Trident missiles were thirty-four feet high and were carried, vertically, inside the submarine. "Did you ever see a Trident submarine?" said Drell. "It's huge. It's really big. Skyscraper."

The air force's MX missile was two times bigger, seventy-one feet high. Garwin and Drell's Jason report said that the most sensible way to base such large missiles was to attach them horizontally to the outsides of small submarines. "And instead of having these submarines with the missiles stuck sideways in them, or vertically," said Garwin, "you would have these missiles lying down in their capsules next to the submarines, external." If the missiles needed to be fired, they'd separate from the submarine, rotate up under the force of the firing, and take off. Garwin said the little submarines with their big missiles were called SUM, for small undersea mobile system; in the Jason report's title, SUM stood for "smallsub underwater mobile." In addition to Garwin and Drell, the coauthors included Freeman Dyson and Bill Press.

SUMs would be mobile, fifty to one hundred of them in the shallow waters along the continental coasts. "Have a lot of little submarines running around with a couple of missiles each," said Drell, "and they would be too many targets." They'd be difficult to find and, Drell said, "they'd be survivable." "Survivable," in the

missile basing debate, meant not that people or cities would sur-
vive but that the warheads would. SUMs would have small crews of
around twelve, would be powered by diesel engines, and would
weigh around a thousand tons. "Five hundred tons would have
done it," Garwin said, but the Jasons thought the number too small
to be attractive. The current nuclear-powered subs were closer to
ten thousand tons. While Jasons were working on the study, they
toured a small submarine nearby in the San Diego harbor. It was
only a hundred tons with a crew of six, built to carry tourists around
underneath Lake Geneva; the sub's captain was five feet five
inches tall.

SUM was debated in the press and in Congress, catching the
overflow from the intense and bitter debate over MX basing. Drell
and Garwin testified to Congress, gave talks, and wrote articles. A
defense scientist said that a "Soviet barrage" aimed at coastal waters would create a tidal wave that would crush the SUMs. A Pentagon official said that, given the capacity of American boatyards, the
only way to build small submarines would be to order them from
Germany. Congress's Office of Technology Assessment studied all
missile basing schemes and was, on the whole, in favor of SUMs.
Aviation Week & Space Technology said SUM stood for "shallow
underwater mobile." In a *New Yorker* cartoon, one urbane, woozy
businessman sitting at a bar says to another, "Ordinarily I lean toward a land-based MX system. But when I have a few drinks, I lean
toward those little submarines."

After a while SUM died. Bill Press said it "just dropped into the
pond like a stone and sank." The air force didn't want its MX missiles on navy submarines; the navy didn't want to jeopardize the
future of its own Trident missiles. "MX was an air force missile, so
turning it into a navy system wasn't really going to happen," said
Press. Press said SUM stood for "small underwater missile." Perry

told Garwin, "I'm glad it wasn't the deep underwater missile, or we would have to call it DUM." Garwin said, "The navy didn't want to do it, the air force didn't want to do it. But it's a really good idea."

Perry said the government never did find a good solution to the MX basing problem, and SUM was no worse than any of the others: "We finally solved the problem by having the end of the Cold War," he said. "MX is now based in warehouses, dismantled. That's the best place for it." Press and Drell thought that Jason's MX basing reports had helped put MXs in the warehouses. "The debate stretched the problem out, and eventually a wiser result came out of it," Drell said. "So it was a worthwhile debate." On the wall in Drell's office is the *New Yorker* cartoon about the little submarines. On a shelf sits a bright toy submarine with little missiles attached to its sides; his students gave it to him, he said.

Jasons love studies, like the one on SUM, in which they invent things. They cite these studies unasked. Their voices change; they sound less like wary government advisers and more like scientists powered by small, barely contained explosions. Like Garwin, they think their inventions are good ideas, really good ideas. Jasons are also partial to the studies, often highly classified, on scientifically unsound ideas, like the dense pack missile basing system, or the navy's repeated attempts to detect an enemy submarine's undetectable neutrinos. Jasons call these studies lemon detection; a former high-ranking defense official said that giving Jason a lemon detection study was "like throwing raw meat to a lion."

But of necessity Jasons also like studies that are useful, that have an impact; and so Jasons were a little bemused by the SUM study's misjudgment about the willingness of the air force and the navy to share assets. "It was Jason's attempt to think out of the box," Bill Press said, "but it was too far—way too far—out of the box."

For a study to be used, not only does it have to be politically realistic, but someone has to want it in the first place: Ari Patrinos wanted Jason's ideas about the climate measurements; the SUM study got as far as it did partly because William Perry had commissioned it. Perry, from the insider's side of the fence, explained that advisory groups too often ask themselves questions and then are surprised when the sponsor isn't interested. "But if it's his question, if it's a question he's seized with and searching to find an answer for, then he's much more interested," Perry said. "I was in the position all the time I was dealing with the Jasons of being terribly interested in their answers, because I had asked them the questions."

A useful study must also address a question that the advisers are good at answering. Perry knew the Jasons' particular talents. Jasons, using a physics metaphor, call this match between sponsor and Jason impedance-matching. The impedance of, say, a computer cable is a measure of how much it impedes the flow of electric current, therefore a measure of how much current gets through; if the cable is plugged into a second cable and the two are not impedance-matched, the current gets through badly or not at all: no flow of current, no communication. If a lesson was to be found in Lukasik's complaint that the physicist-Jasons couldn't help with DARPA's Internet and materials science problems, it was that Jason and its sponsors needed to be impedance-matched.

Impedance-matching was now a bigger problem than it had been when DARPA was the only sponsor, so Jasons matured bureaucratically and set up a committee of semiofficial liaisons. The so-called program committee was organized loosely and never met. Sometimes several Jasons were liaisons to several sponsors; or sometimes members of the steering committee assigned themselves liaisonships; and the program committee and the liaisons changed with time. Ed Frieman, for instance, was liaison for the navy for a while; then Curt Callan was. Gordon MacDonald was liaison for the CIA for a while, then Will Happer was, then Bill Press. The job of the program committee followed the same rules as those of any good relationship: talk to the sponsor, see what the sponsor's problems were, match those problems to Jason's capabilities, follow up with the sponsor about the usefulness of Jason's results, go visit, ask after the families, get to know each other.

Another Jason invention, done about the same time as the SUM study, is an example of an impedance match that was almost perfect. It began as another problem in Cold War missile defense, resulted in a technology called adaptive optics, and ended by helping to free astronomy from one of its fundamental limits. It was actually several studies with a long history; the technology begun in deepest secrecy was released into science's open and invigorating air; and Jasons are unreservedly pleased and proud about the whole thing.

In 1972 Rich Muller was a new Jason working on his first study. "That first summer was traumatic," he said. "It was intimidating to suddenly be surrounded by so many really smart people who knew so much. But in fact I had a very creative summer." That summer, he thought, he did some of the best work he's ever done in Jason. He was trying to find a way around a problem DARPA had: they wanted to use telescopes to identify freshly launched Soviet spy satellites, but the atmosphere distorted what they saw.

The atmosphere had been giving astronomers this trouble all along. The atmosphere up to and just above ten kilometers is a flowing, turbulent river, some patches colder and denser, some patches hotter and thinner. Rays of light from stars hit different patches of differing densities and bend differently. And because the patches are moving along with the river, the light bends differently over time. A star—or a satellite, or a missile—seen through the atmosphere is not a pure bright dot; it's a jittering, twinkling amoeba. Muller was trying to find a way to resolve that amoeba back into a dot, to remove the image's distortion. One way to remove distortion had been suggested back in 1953 by an astronomer named Horace Babcock: deform the mirror into a shape that was equal but exactly opposite to the distortion in the incoming light, and the image of the star or satellite will look the way it should. Babcock's idea was smart but, given the current technology, not possible; by 1972 astronomers hadn't forgotten it, but they hadn't done much with it, either.

One night that summer Muller was wandering the streets of La Jolla "forming pictures in my head," he said, "and thinking about how one could correct for distortion, how you would know if you'd done the right job. And I came up with a really amazing idea—I'm still amazed that I came up with it." The next day he went into his Jason office and began checking his idea with a mathematical means for figuring out whether the mirror's correction was itself correct. Some of the equations he used were complicated, and though Dyson had earlier shown him how to use these equations, Muller still "got all hung up in them," he said. "I couldn't solve them." So he walked down the hall to Dyson's office, where Dyson and Steven Weinberg happened to be talking about the same problem. Muller explained his amazing idea to them. Weinberg said, "Oh, I doubt very much if that's true." Dyson said, "On the contrary, Steve, I would be very surprised if it weren't."

Weinberg went to the blackboard, wrote down the first equation, "and then he did some manipulations on it," said Muller, "and stood back." Dyson said, "I think if you make a substitution of variables now—." Weinberg said, "Oh yes, of course," and wrote several more lines. "I was taking notes," Muller said, "but I wasn't sure what he was doing." Weinberg paused in his writing, and

Dyson said, "Now evaluate the delta function," and Weinberg said, "Oh, okay." Weinberg wrote a few more lines, and Dyson said, "Good. You've proven it." Muller's idea was right. Weinberg and Dyson, said Muller, "are two of the most incredibly brilliant people I have ever met."

At the end of the summer Muller went back to Berkeley and told his colleagues about the study—it was unclassified—and they "actually built the system that worked," Muller said. That is, the mirror deformed and the star's image sharpened up. They published their idea in a scholarly journal, but in the end, Muller said, his method would work only on bright things and not on the faint stars that astronomers most want to see. That same summer Dyson had done another Jason study, analyzing the effectiveness of all methods of sharpening images, and ended up with another drawback, that no methods would work well with visible light. All methods of image sharpening, Dyson found, worked best with stars that were infrared bright; they still do. Dyson also published his work.

Meanwhile, an entirely different and unconnected part of the scientific universe was working on the same problem. In addition to funding Jason, DARPA was also funding both the aerospace industry and air force laboratories to invent ways of sensing the exact distortion in an image and of correcting that distortion with a mirror. And these industry and military scientists also built systems that worked. In the mid-1970s, on a small DARPA telescope on Maui at the top of Mount Haleakala—which, at three thousand meters, was as close to the top of the atmosphere as they could get and still be on the ground—they installed what was now called an adaptive optics system. Over the next few years, the industry and military scientists figured out how to deform the mirror more precisely and finely. Calculating the image's exact distortion was still possible only if the image was bright, but at first this was less of a problem for the military because it didn't want to look at stars; it wanted to look at, or maybe shoot down, satellites and missiles, which were generally bright.

But then the industry and military scientists decided they also wanted to see things that were fainter. They had the idea of using a laser like a flashlight to shine a bright dot—an artificial star, called a laser guide star—in the sky, on top of the image of the faint thing. Then they would calculate the distortion in the image of this guide star, correct the mirror accordingly, filter out the laser light, and look at the faint thing. In the early 1980s, when the industry and military scientists were trying to work out how to do this, they were only a handful in number and they still didn't know whether their idea was classified. One of them was Robert Fugate, an air force scientist. He and his colleagues knew "what great benefit this would have to the astronomy community," he said, but they also knew not to discuss their idea openly. In any case—even though "guide star" is an astronomical term for a bright star used as a place-marker for a nearby dim one—astronomers didn't know about these systems.

In June 1982 Fugate asked Jason for help on a specific problem. The laser guide star faded out with height, as a flashlight does; it didn't go high enough into the turbulent atmosphere, and images of anything higher would remain uncorrected. Moreover, the laser guide star worked best on small telescopes, and industry and military scientists wanted the systems on larger telescopes, which could see fainter things farther away. Fugate and David Fried, an unaffiliated expert on atmospheric turbulence, went to the Jasons' summer study and presented their problem. Will Happer said, "There's this nice layer of sodium up around ninety to a hundred kilometers, and all you have to do is shine a laser on it and you can make an artificial star." Fugate said, "And David and I looked at one another with dumbfounded looks on our faces like, 'Is there sodium at ninety kilometers?'"

In the mesosphere, near the top of the atmosphere, is a layer of sodium atoms deposited by meteors that hit the atmosphere and vaporize. Happer said he knew about the sodium layer by chance: "I don't know why I knew it, but I knew it." Happer was expert in making atoms shine by tickling them with lasers. He also knew that lasers can be tuned to varying colors; a sodium laser is tuned to make the sodium atoms in the mesospheric layer fluoresce an orangey-yellow, like a sodium streetlight. And since that sodium layer was ninety to one hundred kilometers up, it was above the lower turbulence, and effectively all the distortion could be corrected. Happer, working with Gordon MacDonald, spent the rest of the summer figuring out whether the atmosphere's sodium layer had enough sodium to fluoresce brightly enough to make a nice bright guide star, and luckily it did.

The sponsors "were very happy," Happer said. "They were so excited about this that they clamped down really hard on security." The new idea became so completely classified in such a small compartment with such a short list of those who'd need to know that for a while even Happer and MacDonald didn't have the clearances to know about it. "DARPA said that since we don't really know the ramifications of all of this," Fugate said, "we're going to keep it classified until we know more." Happer and MacDonald were a little unhappy that something so useful to astronomers was a secret, but they didn't do anything about it.

To learn more, DARPA funded the industry and military scientists to run two experiments. One, on the short-range flashlight laser guide star, called a Rayleigh guide star, was run by Fugate at the Starfire Optical Range at Kirtland Air Force Base. The other, on the long-range sodium laser guide star, was done at MIT's Lincoln

Laboratory. A year later, in mid-1983, Fugate's Rayleigh guide star system passed its proof-of-principle test; and a year after that the sodium guide star did the same.

Meanwhile Ronald Reagan had become president, and though PSAC had never been resurrected, a new and smaller science advisory committee did advise the president, though it didn't report to him directly. On March 23, 1983, Reagan announced a new missile defense program that he called the Strategic Defense Initiative, or SDI, and that its critics called Star Wars. SDI's plan was to locate the country's missile defense system out in space, thereby providing an impermeable antimissile umbrella over the whole country. Reagan's science advisory committee was surprised; they hadn't heard of it before. Over the next years Jason reviewed SDI repeatedly; one Jason felt they were pushing the limits of how much independent and often negative advice the Defense Department would continue to pay for.

Though concrete examples of the SDI umbrella were sparse at the time, they included high-powered lasers and charged-particle beams stationed out in space to destroy any enemy missiles. Careful aiming would be crucial, and the sodium laser guide star was obviously relevant, as was adaptive optics. In 1984 all research was moved from DARPA into SDI. "When SDI was created," Fugate said, "they just went around and scooped up everything they thought would apply, and this was one of them."

Jason had continued working on adaptive optics and the sodium laser guide star; Happer and MacDonald were joined by several others, including Dyson and Claire Max. The Jasons wrote three reports, in 1982, 1983, and 1984, Max said—"that was when we really did the work." The Jasons were also still thinking about astronomy, and for the 1984 report, Max said, "Freeman Dyson and I went around and talked to the astronomers at UC San Diego, and put our heads together," then worked out the design of an adaptive optics system mounted on a telescope along with a sodium laser guide star. Max and Dyson suggested that this design "would also be terrifically good for astronomy," she said, "in particular, for the ten-meter telescope that was under consideration for what became the Keck Observatory." The report was classified, though, so no one could tell the astronomers.

The tightness of the security makes all the more refreshing the publication in 1985, in the completely open and international journal *Astronomy and Astrophysics*, a paper by two French astronomers suggesting that adaptive optics might be aided by sodium laser guide stars. "That French paper basically described the concept," said Fugate. "And did it very well. And that's all I'll say about that." What Fugate could say, however, was that the French astronomers' publication finally got the attention of the astronomical community.

After the French publication astronomers began sending the National Science Foundation—NSF, the funder of much of the country's academic science—proposals for grants to build laser guide star systems. The person who got the proposals was the director of advanced technologies in NSF's astronomy division, Wayne van Citters. Van Citters also happened to have worked on an SDI advisory panel and therefore knew about the laser guide star experiments at Kirtland and Lincoln Laboratory. So van Citters asked the scientists "on the other side of the curtain," he said, to review the proposals to see whether the astronomers "would be reinventing something, or barking up the wrong tree."

Clearly, the sodium laser guide star was now public and the cat wasn't going to go back into the bag. Van Citters, Fugate, and other scientists—including William Thompson, who was in charge of the air force's adaptive optics programs—began meeting to consider their options for the astronomers' proposals. The first option was to do nothing, to refuse to fund the proposals and let American astronomers struggle along the best they could. One problem with that option was that by now European astronomers had begun installing adaptive optics systems, without laser guide stars, on telescopes and making American astronomers look bad. The other problem was that the military scientists were not only security-minded military but also open-minded scientists, and they wanted to see what improvements the astronomers might think up. "We're researchers at heart," said Fugate. "And we just, you know, liked to see the advancement of technology."

The second option was to fund the astronomers, but the problem with that was obvious: the American government and therefore the taxpayer would be paying for the same expensive technology twice, once through the military and again through NSF. "So that wasn't real attractive, either," Thompson said. They chose the third option: declassify the technology and turn it over to the astronomers. All along, keeping the technology classified had had the drawback of much of classified research: the scope of the military programs in adaptive optics was small enough that "you can't bounce ideas off of people," said Fugate. "You can't get other people's thinking on the subject." He added, "There are a lot of smart people in the astronomy community. And so in our view, there was a lot to be gained by opening this up."

The air force began the painstaking process of declassification by parsing out and setting aside the parts of the technology that would remain classified. Then everyone with a high enough clearance to read the request was asked to balance declassification's costs and benefits. The request went not only to NSF and Lincoln Laboratory but also, said Thompson, to "a dozen or twenty organizations that were working in this field," including various defense offices that might use the technology, to the Defense Intelligence

Agency, possibly to the French intelligence agency, and to Jason. "We did get a response back from the Jasons," which, said Thompson, "is classified."

National security is unlikely to be threatened if I reveal that the Jasons liked the idea of declassification. Claire Max wrote Jason's response, saying generally that "first of all, the astronomical community is discovering this by themselves—why make these people rediscover the wheel, which they were clearly going to do anyway in a few years? Second of all, it would be a tremendous boon to the new generation of giant telescopes that the country is paying a lot of money for. And the astronomical problem was actually easier than shooting lasers at satellites, and therefore, you know, crucial parts wouldn't have to be declassified." Jason's basic argument, she said, was "save some trouble and do this thing that would be really good for astronomy."

In fact, Jason had been pushing for it all along. Max had been lobbying the Pentagon. "Claire Max was the real hero," Happer said. Dyson said the same. Max doesn't say she's a hero, but she does say she talked a lot about declassification with Fugate and also "took every occasion I could to talk with DoD people when they attended Jason meetings in both La Jolla and D.C., where we could talk in a classified environment."

For around two years the request to declassify the laser guide star went up and down and back up the defense hierarchy. Finally, on May 23, 1991, the air force opened the technology to astronomers, announcing it in a press conference at the Pentagon. On the same day Fugate and a scientist from Lincoln lab gave a scientific talk at the spring conference of the American Astronomical Society. The talk's title—previously cleared by the Pentagon for public release—had been announced in the widely distributed bulletin for the conference, Fugate said, "so there was quite a lot of anticipation at that meeting, and I was kind of overwhelmed. There were like six hundred people in the room, and I hadn't talked to a group bigger than thirty people on this topic before."

Fugate told them the air force had a working adaptive optics system with a laser guide star on a small telescope and showed some pretty pictures of the results. The audience of astronomers had mixed reactions—surprise, curiosity, and anger. One of the speakers after Fugate was an astronomer who had spent years trying to verify whether a sodium laser guide star would ever work. "So he got up," said Fugate, "and he said, 'I've never been so scooped in my entire life. I think I should just sit down.'" Max said the astronomer who said he'd been scooped ended up being "really quite angry. I can imagine how awful it must feel to find out that people knew about it all the time and they weren't telling him. You know, that's not good." Fugate said that afterward the astronomers had "a million questions, just every variety you could imagine: 'Where can I get one of these? How does this really work? How much does it cost? Can I put this on my telescope?' It was just pandemonium."

Charles Townes happened to have been chairing the conference session at which Fugate talked, and not by accident. Townes had learned about adaptive optics and the sodium laser guide star through Jason. He had long since won the Nobel Prize for inventing the principle behind the laser, and he was now doing astronomy. In need of sharp images, he had begun to drop in on Fugate. Convinced that Fugate's work had been "outstanding and forefront" and should be publicly recognized, he persuaded those in charge of the astronomical meeting to invite Fugate to talk.

Townes knew about the effort to declassify the system and had done his own personal lobbying "with some high official," he said, toward that end. The rumor was, Fugate said, that the final authority for this declassification was the president, then the first President Bush. "So when it was presented to the president," said Fugate, "the staff person said, 'Charlie Townes thinks this is very important, that we should proceed with it.' And the president's response was 'Well, okay. If Charlie said it was something we need to do, I'll sign it.'" Fugate said that was the story he heard, but he didn't know if it was true. Bill Thompson at the air force lab was dubious: the air force had the "original classification authority here, with our commander," he said. "We did not need to go back to Washington for a signature." Charles Townes was inscrutable: "I remember some such statement about the president, but don't really know what he said anent adaptive optics." I didn't know whether Townes even knew President Bush, but I was too impressed by the word *anent* to ask. Fugate seemed to think that if the story wasn't true, it should have been; he admires Townes greatly. "It's just been one of the highlights of my career," he said, "just being able to say I know him."

Three years later, in 1994, the *Journal of the Optical Society of America* published a two-volume issue explaining the state of the known art of adaptive optics. The issue included a declassified, merged version (by Happer, MacDonald, Max, and Dyson) of two of Jason's three reports. The Jason reports themselves, however, are still classified, and when I asked Max what they were about, she said, "I can't tell you what was in these reports." Was one of them maybe about the adaptive optics system and the other about the sodium laser guide star? "I can't tell you," Max said. "I know, and I can't tell you."

Dyson and MacDonald both were annoyed about the decade during which the adaptive optics/laser guide star system was hidden in the classified literature. "I think it is not an exaggeration to say that the secrecy held up progress in adaptive optics for ten years," Dyson said. Max thought Fugate and Lincoln lab were exceptions to Dyson's criticism but explained that once a military technology becomes "operational," military researchers are supposed to stop improving the technology and just operate it. "The people inside the secrecy barrier," Dyson said, were happy with the system they built and stopped looking for ways to improve it. "The people outside the barrier did almost nothing because they wrongly imagined that the people inside were pushing ahead. The only people who actually pushed ahead during those years were the French."

The other person who pushed ahead was Claire Max. Max was an unusual Jason, partly because she was the first woman Jason and partly because she was not an academic. She got a Ph.D. in physics from Princeton, then went to the Lawrence Livermore National Laboratory, where she worked on how laser light interacts with hot, charged gases called plasmas. When she was invited to Jason she said she "was looking for some variety in my professional life." She remembered having had the impression that Jason "was a congenial group of people," she said, "and it was exciting and stimulating. That's still what I would say, I think." She had joined Jason in time to work on the second and third adaptive optics studies.

Over the next few years, after working on those studies and reading the French astronomers' publication and other articles on adaptive optics by American astronomers, she said, "I had thought that this idea of sodium laser guide stars was so good for astronomy that surely somebody would pick up the ball and run with it." But by the early 1990s only Lincoln lab had a sodium laser guide star, still in the experimental stages. It was, Max said, "a giant, expensive, complicated system. It wasn't the kind of thing that you'd want on an operational telescope. It was a beautiful experiment, but it was too much of a heroic act." Max thought to herself, "Oh, this is silly, nobody's done this, Somebody's got to do it. And I'm at Livermore where they do lasers." Shortly after declassification and Fugate's talk to the astronomers Max used Livermore funding to "shine one of these lasers up into the sky and actually measure the turbulence," she said. "And it was an amazing experiment."

In particular, Livermore had lasers that had been built to separate out fissionable uranium for civilian nuclear power plants and that could be adapted for telescopes. The lasers were in one building, and their light shone into another building where the uranium was separated, through an underground tunnel. At one place in the underground tunnel was a manhole from the tunnel up to the ground, and on the ground was a faceplate. Max and her colleagues pulled off the faceplate and inserted into the tunnel a mirror angled so that a laser light racing through the tunnel would hit it and bounce up into the sky. Then they bought a laser-aimer from a telescope company; installed it over the manhole; protected it by a big yellow tent with a fold-back flap; and inside the tent, set up a little telescope. Then they got the local powers-that-be to turn off the streetlights; and with a mirror, laser-aimer, tent, and little telescope (not counting the Livermore laser), they proved that sodium laser guide stars would allow them to measure turbulence. "It was preposterous enough that we couldn't believe that we were doing it," she said. "But in fact it worked really beautifully. It was a real gas, this whole thing." Max has a photo of a glowing yellow tent against a black sky and a laser shining straight up to God. "I still get gripped by it," she said.

What Max likes about the history of the sodium laser guide star is the unusual way the military and the astronomers have leapfrogged. "First the military advances something," she said, "and then the astronomers do, and then the military does, and then the astronomers do. It's like a braid almost." Academic and military scientists generally stay at arm's length, partly because of classification, partly because as pure and applied scientists their problems are often different, and partly because they're at different levels in the professional hierarchy. Max and Fugate both said the braiding continues, that the two formerly noncommunicating cultures have good relations, that they go to each other's conferences, that people who work on adaptive optics for the air force have moved over into the academic community. Fugate, whose military community was relatively small and secretive, said that before he gave that talk to the American Astronomical Society, he hadn't spent much time with astronomers: "I've never run into a more closely knit, well-networked, everybody-knows-what-everybody's-doing kind of thing and everybody is willing to help everybody. It's a great group of people."

The adaptive optics and sodium laser guide star studies were some of the few unrelievedly good results of SDI, a program that scientists agreed was generally useless. They "gave Jason quite a bit of cachet for a few years," said Happer. Dyson said it was the most serious work he's ever done in Jason. "That was great fun and was a serious science problem," he said, "and I think we did some really good work on it."

By 1989 SDI had run into serious problems. Politicians and press alike criticized the program for jeopardizing the 1972 ABM treaty and destabilizing the Cold War standoff. Scientists said it didn't square with the laws of physics. Physicists' largest professional organization, the American Physical Society (APS), said years more of research would be needed before anyone could decide whether space lasers and particle beams were even feasible; the president of the APS at the time happened to be Sid Drell, and "no," he said, "it was not coincidental."

So Reagan's successor, the first President Bush, decided to drop space lasers for a program invented at Livermore called Brilliant Pebbles. Clouds of small satellites, each fitted with infrared sensors, computers, and little rocket motors, would detect and home in on enemy missiles, then ram the missiles and smack them to pieces. In the summer of 1989 the SDI office asked Jason to review its Brilliant Pebbles program.

Jasons were now doing increasing numbers of reviews of sponsor's programs, like those done for Ari Patrinos's climate program. Though they knew that program reviews were often valuable, they weren't crazy about doing them. Program reviews rarely involve inventions, and by the time Jasons were asked for a review, the program had sometimes already been invested with sponsor's resources and loyalty. So the review's value was mostly political, for or against. From Jason's point of view, program reviews were too often bad impedance matches.

Jason's review of Brilliant Pebbles had nineteen authors, was three volumes long, and was classified. The following February SDI's director released an unclassified summary he'd requested of the review and announced that Jason had "endorsed the concept"; he therefore planned to triple Brilliant Pebbles' budget, to $55 billion. The press sent out reporters. Jasons, looking for the fine line between protecting the sponsor and telling the truth, told the press that the director was overstating. Brilliant Pebbles as a defense, said one Jason, would be leaky at best and its technology needed work. The director's office said that the study expressed confidence that any problems with the technology could be fixed. Happer, who was a coauthor, told the press that he didn't remember expressing confidence that problems could be fixed. "Maybe they can and maybe they can't," he said. He said that countermeasures were plentiful and cheap, and he'd hate to be asked to guess the eventual cost. "I don't feel it's all a crock," he said, and in fact he thought the program had a lot going for it—which was saying a lot, he added, given that SDI's earlier space lasers were clearly "a stupid idea." Afterward Happer worried that because he was Jason's chair, his bluntness might "cause Jason itself to get the ax," so he talked to the SDI office and added some weasel words. "They were very angry at me and at Jason," he said, "but it was not at that point a fatal anger." Some Jasons were irritated that he hadn't stuck to his guns.

In spite of the near-universal criticism of SDI, it turned out to be a bargaining chip in the ongoing negotiations between the United States and the Soviet Union. In July 1991 the countries negotiated the first Strategic Arms Reduction Treaty, agreeing to reduce not only the numbers of missiles but the numbers of warheads. Relations between the countries warmed. The Soviet Union was having internal problems. The following December the component countries of the Soviet Union declared their independence, the union dissolved, and the Cold War was over. In 1993 the new Clinton administration officially canceled Brilliant Pebbles.

With the end of the Cold War, Jason again asked itself, "Whither Jason?" Jasons noted that though they now had around ten different sponsors, not all related to defense, and though the Soviet Union had collapsed, no one thought defense studies were going to go away. They worried about the chaos in the former Soviet Union and about who now owned its nuclear weapons. They noted that worldwide the number of countries that possessed or were capable of possessing nuclear weapons was only growing. "The world is changing so rapidly," Curt Callan said. "Are we going to have ten more nuclear powers?" They thought the military might now face different kinds of wars—like wars in cities, wars in jungles, wars not against nations but against national parties—in which good intelligence and fast communication were unusually important.

And sure enough, within two years of the demise of Brilliant Pebbles, Jason began studying the urban battlefield. Urban warfare is fought not on battlegrounds but in cities, where soldiers "roam around in small groups and get sniped at," said Paul Horowitz.

Cities have corners soldiers can't see around and buildings they can't see into, said Callan, "and all of them can hide dangers. And so the question was whether you could just put a whole lot of sensors out there that communicate to each other and to the soldiers moving around and to the commanders in some useful way." To power all this sensing and communicating, soldiers carry forty pounds of batteries; lighter power sources would be welcome. In general, the urban battlefield presents many of the same problems as the Vietnam barrier but with modern imagery, miniaturization, power storage, autonomy, and bandwidth. "There are questions about communications and imaging and networks and electronics storage," Horowitz said. "Should we be using drones? Should we be using helmet-mounted video cameras? And what do you do with all that data? That involves data fusion, huge amounts of data, artificial intelligence, target recognition."

These post–Cold War kinds of warfare, Jasons worried, would lead Jason into studies that, while unqualifiedly necessary, would be in the long run less important. During the Cold War the country had had two clear central problems: missile defense and nuclear test bans. Jason's role in the Cold War had been likewise clear. "The underlying physics of ballistic missile defense, let's say," Callan said, "was a uniquely identifiable superquestion." And because Jason was expert in these superquestions, Callan said, "what we said would be listened to by the secretary of defense—perhaps dismissed, but we would be heard."

The new world order, after the collapse of the Soviet Union, didn't seem as clear. "The initial euphoric thoughts about how the world was going to be a lot better place turned out to be optimistic," he said. "And we haven't found out what the new order is yet." So far, Callan said, with the different kinds of wars, "it is a hell of a lot harder to identify questions that had the same importance. It's a much more diffuse problem." So Jason might no longer be able to write the kind of technical reports that, Callan said, would "be read by the secretary of defense and cause his blood pressure to rise and throw it in the wastebasket, or alternatively to say"—and here Callan imitated a deep, gravelly voice of command—" 'Jesus, why didn't you guys think of that?' Now it's kaleidoscopic. It's not—" He ran out of words and started over. "Nobody really knows—" He ran out of words again. Drell said the same thing more succinctly: "Put it this way: in the old days there were several clear major problems that focused our attention. Now there are lots of cats and dogs. Lots of cats and dogs."

The problem with the cats and dogs is that they're usually short-range and they can risk triviality. "We have to avoid concentrating only on today's need," said Drell. "And I think the danger of becoming a job shop is always with us." A job shop takes on short-term, one-time projects. A job shop's projects depend only on what the customer wants, so its direction is driven by the customer. Jasons use the phrase "job shop" a lot. "You know, you always think the good old days are the best," Murph said. "But we've been subjected to a great deal of pressure—and succumbed to it—to be a job shop, to do a large number of relatively small things." The steering committee worries about the job shop problem, too. "We've always debated it," Drell said. "We're always questioning it."

Murph worried that small, disparate studies would dissipate Jason's "team spirit" and "fracture the group." These studies might also pose deeper threats: a job shop is the antithesis of the nearly-missionary, genie-corking reasons that motivated Jasons in the first place. Hotshot academic scientists are not likely to spend their summers working in a job shop.

CHAPTER EIGHT

Blue Collars, White Collars

"If you want to keep working from within, you have to maintain the trust of the people who are asking you those questions. And trust means a certain degree of circumspection in what you do and don't talk about in other venues. But I do think that's a tricky one."

—Paul Horowitz, interview, February 15, 2002

Any anthropologist will tell you that to understand a tribe, you must study its stories. This story is about the differences between the worlds of scientists and science advisers; the Jasons tell it unasked and with marginal accuracy.

In 1969 Richard Garwin began a rare second term on the President's Science Advisory Committee, PSAC. One of the science-related issues then before the nation and PSAC was whether to build a commercial airplane that would fly faster than the speed of sound, the supersonic transport, called the SST, which the French and British were already building. The Nixon administration commissioned several ad hoc panels to study the feasibility of the SST; one panel was under the White House's Office of Science and Technology, and Garwin was in charge of it. In March 1969 Garwin's report—like the report of at least one other panel—recommended against the SST; Garwin's recommended that the government drop the whole program. However, Garwin said, "this was not the answer President Nixon wanted," and the following September, Nixon announced that the SST would go ahead.

Meanwhile, the Garwin report had not been made public. Both Congress and the press knew about the report—the Nixon administration had publicly announced its commissioning—and neither one liked being kept in the dark. A year after the report the March 13, 1970, *New York Times* announced, "Report on SST Said to be Kept Secret." At the same time Congress asked Garwin to testify, and he did so several times, in April, May, and August 1970. He told Congress that the SST would be as noisy as "some fifty 747's taking off simultaneously" and would drastically exceed federal noise laws. He said that the administration wasn't being candid with Congress and that the SST being developed was not the one that Congress had approved. Private industry wouldn't risk funding the whole program, he said, so the taxpayer was going to end up with an enormous bill. "I recommend the immediate termination of the U.S. Government's direct or indirect support of the SST program," he said. For this and other reasons, a year and half later in August 1971, urged on by economists, environmentalists, presidential advisers, politicians, and PSAC, Congress killed the SST.

The Nixon administration was unusually convinced of the value of unwavering political loyalty and moreover was fed up with a PSAC that thought it should be politically nonpartisan and that opposed Nixon's missile defense program. The death of the SST was the final affront. "Who in the hell do those science bastards think they are?" said a White House staffer. In late 1972, at the end of Nixon's first term in office, PSAC members submitted pro forma end-of-term resignations; when Nixon began his second term, he accepted the resignations as permanent, and PSAC was fired. Though PSAC has been since revived in various administrations under various names, no science advisory board ever again re-gained PSAC's political presence.

No one who writes about PSAC's firing blames Garwin, but everyone cites him. Whether Garwin was right to testify publicly is a matter of intense argument and little agreement. On the one hand, even though Garwin's report was not for PSAC but for a temporary panel in the White House's science office, and even though Garwin said repeatedly in his testimony that he spoke only for himself and not for the White House, he also happened to be a current member of PSAC. PSAC advises the president from inside the executive branch; its first loyalty is to the president. As such, its advice can be contrary to the president's own opinions and pro-grams, but that advice is private, and any disagreements are kept in house. If PSAC members go public with disagreements, Congress will make political hay of them; and the president might be faced with either holding to a stupid position or changing positions in public, and in any case would certainly lose trust in his science ad-visers. The rule in the world of government insiders is, if a PSAC member feels that public disagreement with the president is neces-sary, the member should quit first and talk later; or as Ed Frieman said, "If you disagree, you either go along or you get out." In 1991, for example, Will Happer, who took a leave from Princeton and Ja-son to direct the Department of Energy's Office of Science, said publicly that Vice President Al Gore's claims of environmentally dangerous amounts of ultraviolet light hitting the ground were based on unreliable measurements, then proposed to make more reliable measurements and was fired. "It was a terrible thing," said Frieman, who had earlier headed the same office, "but it happened because Will was deemed to have not followed the rules."

On the other hand, PSAC members remain citizens with the rights and duties of all citizens. A PSAC member should not reveal what was learned in confidence and should not speak on behalf of the entire committee. And Garwin said repeatedly in interviews, in talks, and in print that he told Congress nothing he had learned on PSAC and nothing that Congress did not already have before them from other sources. He pointed out that the government's official code of ethics begins by saying that a government employee should put loyalty to the country above loyalty to persons, parties, or administrations. He said that the administration, in lobbying for the SST, "concealed relevant information and lied to congressional committees." He said that as far as he was concerned, "had the administration been honest with the Congress, the matter would have stopped there." He'd thought about it a lot since, he said, and thought "perhaps it would have been better had I resigned." But he decided that he would have been equally as newsworthy as a PSAC member or as an ex-PSAC member, and the impact on PSAC would have been the same. When administrations lie, he said, advisers should do their best to "return things to the right track," but when they cannot, he said, "it was much more important for PSAC to have integrity than to simply exist hoping for better days." In that case, he said, "the adviser must act, no longer as an adviser but as a citizen."

PSAC's demise and the pallor of its resurrections have certainly not been what citizens would desire for their country. But citizen versus science adviser arguments aside, the fact is that PSAC's demise probably wasn't caused by Garwin's testimony. Other science advisers to President Nixon testified against administration programs without repercussions: Gordon MacDonald, who was then on the President's Council on Environmental Quality, also testified against the SST and wasn't fired; and when another PSAC member asked whether to resign before testifying against the antiballistic missile program, Nixon himself said no.

Even if the Nixon administration was irked at Garwin, it could have gotten rid of him without killing PSAC—just "toss him off it," said Ed Frieman. "There's a lot of mythology around."

Nevertheless, the story among a number of Jasons is that Garwin "went public" and "destroyed PSAC." They tell the story with detached interest, as though it's ancient history. Part of the reason they tell the story is that telling Garwin stories is a Jason hobby. Garwin abuses admirals. Garwin nails generals when they're vague about details. Before a briefing begins, Garwin walks up to the briefer's table, flips through the briefing charts, and then walks out. Garwin listens to a briefing, does the calculations in his head ahead of the briefer, and ten minutes into the briefing interrupts to say that the calculation is off by three orders of magnitude on the second equation down on the fourth line over. Garwin, said one Jason, is "a gadfly to the military." In Act I of the anniversary party Jason skit, a colonel played by Paul Horowitz sings to the tune of "Just You Wait, Henry Higgins," "Ohhh, Richard Garwin / Just you wait until your clearance is reviewed. / Ohhh, Richard Garwin / Do you really think it's gonna be renewed?" Jasons say Garwin "has a powerful and original mind"; he has "no mush in his mouth"; he's "one of the hardest workers in Jason; he's "the most informed person in the United States on defense problems"; he's a "phenomenon"; he's "sort of a singularity out there." A Jason joke: Garwin is standing in line to be guillotined; the person ahead of him kneels at the block, the guillotine doesn't release, the executioner says it's the will of God, the person goes free; Garwin kneels at the block, the guillotine doesn't release again; Garwin looks up and says, "Oh, I see what the problem is." It's an old joke about techies, but Jasons say "it is *so* Dick," and "it's the continuing acting-out of the Dick Garwin joke by Dick Garwin that is just such pleasure." Garwin likes precision and accuracy and sends messages like this: "I am traveling but will be home Sunday evening, 11/16/03. Call me

anytime after 8 p.m. In the meantime, 'orbiting' has only one 't.'"
In my own experience he is remote but helpful and charming
when you least expect it. And if your thought processes are a little
slovenly, let alone dishonest, he's the concentrated essence of someone you'll want to avoid.

Garwin's account of himself is simpler. "I like technical work,"
he said, and if you advise the government on what will and won't
work, "it really helps to be able to do the calculations yourself,
to be intimately familiar with the technology. So when you are
the best person for the job, and it's a job that should be done"—
he pauses and shrugs—"you do it. So that's what I do. I like to be
helpful. I believe in progress. And I'm very good at taking the next
step."

The other reason Jasons tell the story about Garwin, the SST, and
PSAC is that it's a fine example of the cultural problems that come
up when scientific outsiders try to work with government insiders.
The first problem is whether Jasons should ever take a public stand
on studies they've done for the government. Those rules about going public are fairly clear for advisers inside the government. The
rules are looser for science advisers who are outsiders and on panels for the government or for the National Academies and who are
not paid. The rules for science advisers, like Jason, who are outsiders but are paid are somewhere between clear and loose; these
rules require, said Frieman, "a certain politesse."

Jasons are never told they should not go public. But some Jasons agree with the government insiders' rules and think that scientists who go public on the subjects of their advice are no longer
seen as objective and are therefore less credible. "I think once you
come out publicly for something," said Rich Muller, "you're basically taking sides, and your value as an adviser is decreased." Paul
Horowitz takes Muller's argument farther. "It's not a freedom of

speech issue and it's not a classification issue," he said, "it's a trust issue." The sponsor, said Horowitz, doesn't "want to see in *The New York Times* or *The Washington Post* conclusions that basically short-circuit that trust and say that it is running a bankrupt program." If the program is truly bankrupt, Horowitz said, a scientist should tell the sponsor, not the newspapers or Congress. "I don't mean to put any value judgment on anybody who does that," he said. "But to me, that's too delicate a thing for me to figure out how to do right. And so I have a simplistic approach to this. Which is, we work for our sponsors, we hope they'll do the right thing." If the sponsor doesn't do the right thing, Horowitz said, "Congress will get a whiff of it, one way or another, and they'll put their dogs onto this thing, and that's the way the three branches of government are supposed to work."

Jasons cite examples of the adverse effects of going public, the foremost being the time in the late 1970s when Jason was excluded from working on the technology that makes military airplanes invisible to radar, called Stealth. Jasons variously surmised that the reason for their exclusion was fear that they might make it public or that they were just too meddlesome or that they had annoyed the wrong generals; and reading "Garwin" for "they" wouldn't be entirely wrong. In this case, Jasons' surmise seems to be generally incorrect. The person who could have asked Jason to work on Stealth technology was William Perry, who was then the director of defense research and engineering. He said no conscious decision was made to exclude Jason. "I think using the word 'excluded from' is probably not right," he said. "They were not 'included,' that is, we did not think it was necessary." And was he worried that Jasons might go public? "No," he said. "No. In fact, I will be more explicit than that. I never had a concern about technology or secrets that I shared with the Jasons. I felt they were a very trustworthy group."

Other Jasons—Garwin, obviously, and Sid Drell—think that as long as science advisers reveal nothing privileged and speak only for themselves, they are uniquely independent, uniquely informed, and therefore uniquely valuable. "You are there to help the government, but you're not part of the government, and you're free," said Drell. "And so it gives you both an opportunity and a responsibility to speak out." Drell thinks speaking out must be done "in a way that doesn't make you ineffective because you're just one of the leading mouths on all issues. And some people could say I've spoken out too much, I won't judge that. We have to judge where and how we can use the fact that we really know what's going on. And they can't shut us up. The people inside are trapped. They're part of the system. We're not part of the system."

For Charles Townes, going public was a way of sharing moral responsibility. He said that for most of his Jason work, he could feel he was "helping out and doing good things and seeing that things are safer and better. And if there's something that you think is bad, well, okay, you have nothing to do with it. Or you tell people it's bad. You wouldn't yourself feel bad about it if you're telling other people it's bad."

The next problem raised by the Garwin-PSAC story is the balance that Jasons must find between advising about science and advising about policy—for example, giving advice about how to disguise an airplane so that radar has a tough time picking it up, versus giving advice about whether to buy B-1s or Stealth. As with other contentious issues, the balance between science and policy advice is not something Jason decrees; Jasons are left to find their own individual balances. No Jason thinks the balance should tip much toward policy advice. Some Jasons think Jason should give only science advice: you can trust your objectivity about science and you can't about policy. Jasons aren't qualified to deal with policy issues, Horowitz said: "That's not our field of expertise." If you want to have a voice in policy, he said, "citizens do that through political participation. And Jasons do those things as citizens. But we don't do them as Jasons." Drell would tip the balance more: his reason to stay in Jason—beginning with his first "entrapment"— has always been to work on science problems with implications for policy. "I'm going to do a straight technical problem, I'd rather stay home and do particle physics," he said. "But if I can find a technical problem where I think that the policy implications for national security are important, to me that's Jason work."

All these Jasons go on to say that finding the balance between science and policy advice isn't always a straight shot. Drell says that what he likes to stay clear about is that some problems are not simply technical. "It was a technical problem to put a man on the moon," he said. "And that's because the moon couldn't put out de-coy moons, it couldn't put out its lights, it couldn't run away. That was a technical challenge." Problems like missile defense, however, are also "human problems," in which a superb technical solution by one country will certainly lead to an even more superb technical solution by another country. "Because if we build a system, they'll build countermeasures, and we'll build counter-countermeasures," Drell said, and then pointed out the obvious: "And what you want to be very careful of is not to make the fallacy of the last move." Will Happer said he'd like to think "that Jason tries to call the shots as they see it. Let somebody else worry about the politics. But in re-ality usually you're aware of politics."

The best example of balancing science with policy is a series of Jason reports, all done for the Department of Energy, on a program called stockpile stewardship that maintains the nation's stockpile of aging nuclear bombs. "Certainly stockpile stewardship is fraught with politics," said Happer. "It's a program designed to preserve the Comprehensive Test Ban Treaty. And so there are strong emotions there. Some Jasons—I think particularily of Sid Drell, his whole life has been to try to put the genie back in the bottle, you know, with nuclear weapons."

The cluster within Jason that deals with this genie is the arms controllers. Though younger Jasons joined the arms controllers, its core was the older Jasons, including Drell, Garwin, and Dyson. The Jason arms controllers are actually the local chapter of a national group—"not just a clique within Jason," said Bill Press, "but a whole generation of scientists." They're the scientists who are going to keep nuclear weapons under control even if they have to do it with their bare hands.

In 1963 the nuclear family signed the Limited Test Ban Treaty banning nuclear tests in the sky and under water; by 1974 it was able to sign the Threshold Test Ban Treaty to ban all underground tests over 150 kilotons, roughly ten times the Hiroshima bomb. In the 1980s, the family members were still dithering about whether, having signed the Threshold Test Ban Treaty, they might actually act on it; meanwhile they were still free to continue underground testing. In the early 1990s the family agreed to a mutual moratorium on underground testing in hopes that the next treaty might be truly comprehensive and ban everything, all tests anywhere of any kind.

While all this was going on, in the United States the focus of national worry moved from how to verify compliance with the test bans, to whether all those nuclear weapons in our stockpile might have gotten old and ineffectual. "Young, newly built bombs may have infant mortality," said Garwin, but with "geriatric ones, things can go wrong." A few idealistic voices—like Freeman Dyson's— suggested that nonworking nuclear weapons might be the solution to everyone's problems. More realistic voices answered that one-sided disarmament in an armed world was bad policy, and that mutually assured destruction (with the famous acronym MAD) had prevented nuclear war for over fifty years. So the question was, how would you know you could mutually assure destruction unless you tested the weapons? And if you needed to test the weapons, how could you sign a Comprehensive Test Ban Treaty?

The arms controller Jasons had done scattered studies on every stage of the test ban arguments: on seismic verification of underground tests in the 1960s, and on other signatures of nuclear explosions in the 1970s and 1980s. Throughout the 1990s Jason's test ban studies escalated, stimulated by the Bush *père* and Clinton administrations' efforts to stop nuclear tests. By June 1995 the Department of Defense was proposing that the Comprehensive Test Ban Treaty allow tests whose yield was low, far below a kiloton, arguing that these small tests were necessary to judge the reliability of the stockpile. Other federal agencies were arguing that a test ban treaty allowing small tests was hardly comprehensive. Moreover, these agencies said, because tests for the reliability of existing weapons could not be distinguished from tests of new ones, anyone with money and expertise could build and test new bombs. So in the summer of 1995 Sid Drell led a Jason study of what might be learned from tests of differing yields. Going into the study, Drell had doubts. In spite of being adamantly and publicly opposed to the proliferation of nuclear weapons and therefore of their tests, he thought that small tests might be necessary before the stockpile could be judged workable.

The Jasons on the study included four non-Jasons who were nuclear weapons scientists from the three national weapons laboratories—Livermore, Los Alamos, and Sandia—and who had actually helped design the modern stockpile. By July 16, the fiftieth anniversary of the Trinity test, the Jasons had written the summary-and-conclusions section of a report called "Nuclear

Testing." They based the report, they wrote, "on understanding gained from 50 years of experience and analysis of more than 1000 nuclear tests."

The summary said that to learn anything useful about the long-term health of a nuclear weapon, you had to have a continuing program to test the weapon's so-called primary stage. This primary stage is a low-yield explosion that is "boosted" to trigger the much higher-yield secondary stage of a nuclear bomb. The Jasons found that any test below the boost stage wouldn't test the primary and would therefore be useless. In the summary's language, "Underground testing of nuclear weapons at any yield level below that required to initiate boosting is of limited value to the United States."

The Jason study ruled out a scientific argument for needing to test anything that approached a critical mass and began a chain reaction. It specifically ruled out the need for tiny hydronuclear tests with yields less than four pounds of TNT. "We can maintain our warheads to be as reliable as they ever were," said Garwin, one of the authors, "indefinitely, without nuclear testing." The only tests necessary for the health of the stockpile were the so-called subcritical tests, tests that stop just before the chain reaction begins: no chain reaction, Drell said, "is the definition of zero," as in zero-yield tests. "So," said Drell, "that put a prick in the balloon that said low-yield testing was useful, and the air rushed out." The existing stockpile could be judged safe without testing, and the Comprehensive Test Ban Treaty was signable.

The report itself was classified, but over the next two weeks Drell was invited to brief the interested parties—government and military people with the proper clearances—about Jason's conclusions. Drell, along with Robert Peurifoy, a weapons scientist from Sandia National Laboratory who was one of the report's fourteen authors, briefed General John Shalikashvili, the chair of the Joint

Chiefs of Staff. "Peurifoy and I sat there with John Shalikashvili for an hour. It's what you like to see in government, a man who knew what the hell he was talking," Drell said, "who knew what the important questions were, who could understand your answers and argue back and forth." Drell and Peurifoy also briefed William Perry, now the secretary of defense, along with the secretary of energy, and the upper echelons of the State Department, the Arms Control and Disarmament Agency, the president's science adviser, and members of the National Security Council. "Those are the people I briefed," Drell said. "Didn't see Congress."

Congress, once again, found out about it anyway. The Senate had been arguing about whether to fund the four-pound hydronuclear tests that Jason found unnecessary, and on August 4, 1995, subsequent senators on opposite sides quoted Jason as being both for and against hydronuclear testing. Their confusion is forgivable: the Jason report's summary picked its words carefully, and much of what it said was in the logic of its arguments, so that its conclusions are clear only if you already knew the issues and had read closely. So Drell had to call the senators and sort them out about what Jason did and didn't say. The report's summary and conclusions were read into the *Congressional Record*. The senators voted to keep the unnecessary hydronuclear tests in the budget anyhow. The newspapers didn't let the debate drop: on August 9 *The Washington Post* announced, "Physicists Say Small Nuclear Tests Backed by Senate Are Unnecessary."

William Perry was more impressed by the Jasons than the Senate had been. "I distinctly remember the final briefing I got from them on the Comprehensive Test Ban Treaty," he said. Like Drell, Perry had doubted that he could recommend that the president sign the Comprehensive Test Ban Treaty unless it allowed low-yield tests, which Perry thought would be a "crippling restriction." That restriction, Perry said, "would not have been an important technical difference. but it would have been a huge political differ-ence. It might have been a deal-breaker, it might have killed the whole treaty. Now I was wanting to change my mind. But I went into the briefing not believing that I could make that recommen-dation. And I left the briefing saying, 'Yes, I can make that recom-mendation. I've got enough.' " Perry added that outsiders only rarely make recommendations on important policy issues and ac-tually "turn the tide."

The question of whether the stockpile needed low-yield testing, Perry said, "was a physics question" and therefore a "natural" for people with the background and expertise of the Jasons. His only worry going in, he said, had been what Stephen Lukasik's had been, that the pro–test ban Jasons might bias their advice, "that you might end up with people who had the right expertise but who had an ax to grind." What decided him was his "great confi-dence in the competence and the integrity of the particular Jasons who were giving the briefing to me," he said. "I mention Sid in par-ticular because he stands out in my mind as having worked so hard on that problem."

That's a compliment to all the Jason coauthors but especially to Drell, whose devotion to test bans could easily be called "an ax to grind" but whose advice is constrained by the laws of physics. Sci-entists who fudge science to fit their personal politics undermine their own professional foundations. They might as well not be sci-entists. "I would not want, as a Jason, to be involved in studies that are building on a prejudice," Drell said. "Not everybody will agree with my statement—they'll say I am doing that. But I'm just telling you the way I make my own judgments."

The other reason that Perry could recommend a comprehen-sive test ban was that Jason wasn't alone, that the weapons labs had come along. Unlike Jason, the labs had—to quote Happer from an-other context—a dog in the fight. They had responsibility for the stockpile, their jobs depended on it; and continued testing would have made that responsibility more bearable and their jobs more secure. In their own interests, they should not back a comprehensive test ban. "The whole culture of the labs," said Bill Press, who for six years was director of research at Los Alamos, "was that this would be impossible for them to live with." But Jason had been careful to include lab scientists in the study, and the study's report had specified that to maintain a healthy stockpile—"to do the required replenishing, refurbishing, or remanufacturing of age-affected components, and to evaluate the resulting product"— the country also needed to maintain "an experienced cadre of capable scientists and engineers" at the three weapons labs. In the end, the labs backed the report and therefore the comprehensive test ban.

What happened next isn't inspiring. A week after the Jason report, on August 11, 1995, President William Clinton announced that the United States planned to negotiate for a true zero-yield test ban. A year later, on September 24, 1996, Clinton signed the Comprehensive Test Ban Treaty, generally abbreviated as CTBT. In the next days so did seventy other countries, including Britain, China, France, and Russia. By November 1997 fully 148 countries had signed. Yet in the fall of 1999, after much of the rest of the world had ratified the treaty, and in spite of the backing of a large number of generals, admirals, Nobel laureates, and professional organizations, the U.S. Senate refused to ratify it. The senators believed, on balance, that other countries could not be prevented from cheating and that the stockpile could not be judged healthy without testing. Garwin and Drell both testified to Senate committees and held up firmly under pointed questioning by the senators; Garwin even had the Jason summary read into the record again, but to no avail. Happer thought that the Clinton administration, which favored the treaty, had underrated the opposition: they "hadn't done their homework and they felt they could get away with just politicking. And then all of a sudden the Senate holds hearings and the administration finds that most of their supporters vote with the other side. It was a bad show."

Garwin gave a conference talk based on the report shortly after it became public. He said that beginning in 1950, every summer for a decade he had worked on nuclear weapons at Los Alamos; that he had "had a lot to do" with the designs of the first hydrogen bomb and of all its progeny; and therefore, he said, "I am most pleased to be an author of this document." Drell said that in spite of our country's not having ratified the treaty, "we're abiding by it, we're not testing. And the world has signed a CTBT, an honest CTBT. The world isn't testing." That there is a "real, true, honest" CTBT, he said, is attributable to Jason. "There's just no question about that. That's a success worth working ten to twenty years for. Because it matters."

Since the 1995 "Nuclear Testing" report, the Jasons have continued to work on all aspects of stockpile stewardship. Their reports include "Remanufacture," "Subcritical Experiments," and the one Dr. X misinterpreted, called "Signatures of Aging." Another was "System-Level Flight Tests": "The point here is that one wants to test all aspects of the system," Drell explained, "whether the missile will work, whether it'll land in the right place, and then whether it'll trigger the bomb. The delivery system and the warhead and the bomb in the warhead. It's the whole enchilada."

Jason also reviewed the large and expensive programs carrying out stockpile stewardship in the three weapons labs. These programs were the labs' return for supporting the treaty; Happer called the programs "Christmas presents." All three labs collaborate in a large program to take care of stockpile stewardship's supercomputing needs; Los Alamos has a fancy X-ray machine that makes detailed images of bombs as they explode; Sandia has a semiconductor factory; and Livermore has the National Ignition

Facility or NIF, which aims high-energy lasers at a solid pellet of deuterium and tritium, in hopes of achieving a kind of fusion called inertial confinement fusion, or ICF.

Of all the programs, NIF has been the most controversial. NIF has been billed at various times as a way to test the reliability of a nuclear bomb's secondary explosion, and as a prototype for fusion machines for civilian energy needs. Its critics say the secondary has never had a problem with reliability; that fusion for energy is a long way off and seems likely to stay that way; and that given NIF's dubious usefulness, its cost—projected in 2001 to be $3.5 billion—is way out of line. Jason got into the controversy by twice recommending that NIF be supported, once in a 1994 report, "Science-Based Stockpile Stewardship," and again in the 1996 "ICF Review." Jason's reports present the scientific side of the political weapons labs trade-off: NIF would help understand the minute details of fusion, which would in turn help to validate computer programs that simulate tests of nuclear weapons and that obviate the need for real explosive tests. And because the state of matter created by those lasers is also relevant to the fields of materials science, astrophysics, and atomic physics, NIF is also a way of attracting scientists to the national labs. Steven Koonin, a Jason from Caltech who worked on the 1994 report, said, "I had no trouble telling a few of my graduate students to go off in that field, because it is exciting science."

Getting good scientists to the labs—another reason the labs agreed to the treaty—is in no way a trivial issue. Weapons labs charged with an intact, ready stockpile had best not be staffed with the scientists that physicists refer to as "the C students." The only way to bring good scientists into the weapons labs is to offer them the chance to do good science. Drell said that the country relies on the whole stewardship program to "ring a bell if something is going wrong, and this means that the laboratories have to be important places for good scientists to work."

Obviously, by the time Jasons are recommending stockpile stewardship so that the nuclear family can ban tests and the national labs can recruit good scientists, they're a long way from giving purely scientific advice.

Jason balances these insider-outsider issues—scientific advice versus policy advice, the right to testify versus the sponsor's trust—by balancing the Jasons themselves. One early Jason semiseriously divided Jasons into white collars and blue collars. White collars are insiders, or at least have inside connections; they both do science and talk policy. Blue collars are outsiders who just do science. Jasons don't take a white collar seriously unless he is also blue; most Jasons identify themselves as blue; some say that Jasons start off blue and whiten with age; not all Jasons believe the categories exist. Nevertheless, the categories make a point. If the government is to trust Jason's unbiased scientific advice, then most Jasons have to be blue collars. If Jason wants the government to pay attention to its blue-collar advice, then some Jasons have to be white collars.

The white-collar Jasons are part of what William Perry called the "defense scientific advisory community" or the "interlocking directorate of advisers," scientists who regularly populate the government agencies, offices, panels, and commissions. Over the years twelve Jasons—including Drell, Garwin, MacDonald, Frieman, York, Townes, and Murph—have served on PSAC; more have served on PSAC's panels. Both Frieman and Happer directed the Office of Science in the Department of Energy. Press and other Jasons have been on the Defense Science Board. Garwin and other Jasons have been in the Arms Control and Disarmament Agency. MacDonald was on the White House's Council on Environmental

Quality. Drell has been on the President's Foreign Intelligence Advisory Board. "We all serve on jillions of panels for this and that," said Claire Max. The point is, what Jason as an entity does well, or at least well enough to stay in business, is be a blue-collar outfit with enough white collars to be heard.

You'd think that the blue collars who stick to science would act like outsiders and advocate the right to go public; and that the insider white collars would understand the value of loyal silence. But that's not always the case: Jasons regardless of collar go public when and if they are personally inclined to. Paul Horowitz, a blue collar, says that he'd rather work through the sponsors than go public with a criticism. Drell and Garwin, white collars, obviously testify and write about stockpile stewardship. Murph, a white collar, writes articles and signs statements opposing antiballistic missile testing and recent moves by Congress to test small nuclear weapons. So do Steven Weinberg, a blue collar, and Herb York, a white collar. Gordon MacDonald, a white collar, has written about climate policy. Collars somewhere in between write about biological warfare (Steven Block, a Jason from Stanford) and radiological bombs (Steven Koonin). Freeman Dyson, the quintessential blue collar, writes about most of the above. All these issues mix science and policy, and each has been the subject of a Jason study.

Drell explains the balance as though Jasons together make up one person who is thoroughly blue collar but carefully, delicately acts white. Jason has earned respect only, he said, because the sponsors "understand that what we're doing is good technical work and we're not a political game. And I think we'll last only as long as we earn their respect. And that's why it's very important that those of us who do take a more active role in politics don't dominate Jason, don't become what Jason is perceived as. That's a restraint on us."

•   •   •

Drell's reasons for his own white-collar behavior are complex. He has a long, lined face and seen-it-all eyes, and he wears madras sport coats that his friend Andrei Sakharov called his "circus jackets." He talks with energy and force, with a South Jersey accent—"I had nuttin' to do with Jason and Vietnam"—and with the rhythmic, nearly infinite sentences of a storyteller. When he's looking up something, he says quietly, "Doo doo doo doo. Doo doo," until he's found it. He sees complexity in life, and his prerequisites for young Jasons who show promise as white collars are likewise complex: "You look for somebody who's not only a good scientist but has got a balanced view of the world. He knows his limits, he knows his abilities, he doesn't overstate. And he deals with people who are not scientists with respect—he doesn't overpower them, and he doesn't denigrate them. And he knows how to make himself clear." And not to worry about the masculine pronoun, since one of the young Jasons he went on to recommend as a white collar was a woman.

Drell has spent his whole career on the Stanford campus, an arrangement of buildings, colonnades, and plazas in stone and mosaic, terra cotta and ochre, that radiates privilege—power, wealth, assurance, and intelligence. At the gate into Stanford's central plaza is a grouping of Rodin bronzes, the six rich burghers of Calais who gave themselves as hostages for the safety of their city. The message to the inhabitants of Stanford would be hard to miss: with privilege comes responsibility.

Drell's own "social feelings," he said, came from having grown up during the war; from watching the Oppenheimer case unfold; and from being unable to isolate himself and his science from the world that now held nuclear weapons. "And then I came in a generation where the great heroes of modern science had all worked in this area," he said, meaning the area of science advising; and to the extent that he admired his heroes—specifically, Pief Panofsky and Hans Bethe—he felt he should go and do likewise. "I mean, it's important to have models," he said. "And I can tell you, I had *models*."

Nevertheless, he hadn't necessarily wanted to get into science advising. "My own research clearly was far more interesting to me," he said. He was a theoretical physicist, trying to understand the behavior of the particles that made up the atomic nucleus; later he became the deputy director of the Stanford Linear Accelerator, where experiments on those particles were done. "I enjoy teaching and I had my students and I had my children," he said. "So there was no time for things you didn't feel were very important." The importance of Drell's first Jason study, the one on infrared "redout" from an exploding nuclear weapon, lay in its implications for policy—that is, for the feasibility of the MIDAS satellite, and especially for the implications of infrared sensors on satellites. From that study, he said, "I understood that technical intelligence from space was going to be extremely important if we were going to maintain a stable peace in the world." Understanding this, he said, "was crucial and set my whole course."

Drell's course in science advising was limited to two areas, technical intelligence and test bans, which are to him aspects of the same problem, the proliferation of nuclear weapons. Technical intelligence—in this case, mostly spy satellites—meant that "you could talk about arms control based upon verification," Drell said; that is, that you could detect countries that were not complying with test bans. Test bans are "the linchpin of our nonproliferation efforts," he said. "The nonnuclear powers are saying, 'Okay, you big boys, you've had your fun now. If you're going to keep on testing and improving your weapons, we're going to go do it too.' I mean, a CTBT as an end in itself doesn't interest me at all. It's the nonproliferation game."

Without controls on nuclear proliferation, he said, the world has reached a point "where the energies we can release by our actions—the power we have in our hands as a result of technology—can alter the conditions of human survival. That's a fact of our times. We're playing with the conditions of human existence." These sentences are almost theatrical, but Drell doesn't sound as though he enjoys saying them.

And though he sees the details of nuclear weapons as a scientist, he thinks that nuclear weapons are ultimately a moral problem. "We live by values," he said. "They're part of life. You couldn't escape them if you wanted to—they're a nondebatable proposition." The moral problem posed for scientists by nuclear weapons is the same one that the Manhattan Project scientists and the Vietnam Jasons faced, the old problem of *curiositas*, which Drell said was the question of "how the fruits of science will be used."

The decision about the use of those fruits, Drell believes, belongs not to the scientists but to the community. "I mean, you can't control science," he said. "You're discovering something and you have no idea what it's going to be. But the minute you get some idea and you can start thinking about the technical applications, that's where societal questions come in." That is, consideration of the moral questions begins when science suggests applications, and it must be done by society, in a social debate. "And having a debate on these things," Drell said, "that's what I call the moral obligation of the community."

If the community is to consider the morality of science's applications, then obviously the debate has to be public, he said, "not things done in secret," for example, like the decision to develop the hydrogen bomb. "Was there a way that one might have headed off that development?" Drell said. "That decision was never fully debated in public. The ABM decision was. The Nonproliferation Treaty during the Kennedy-Johnson years was. And maintaining the Nonproliferation Treaty now—of which the CTBT is just one small component—that's debated now. And environmental policy, it's being debated now. And I think that's quite essential."

This is where the science adviser comes in, to start the debate and then to inform it; to tell the government and the public about those applications and their potential benefits and threats; to keep the public debate grounded in fact; and to keep it within the bounds of reality.

"This is a point I've always been careful with," Drell said. "The community that generates the new knowledge, that's making the new technology, has a better understanding of its implications. And therefore we as a community have a responsibility to let the leaders of government know how it might be useful, what are the dangers that go with it." In 2002 Drell, a colleague, and Robert Peurifoy wrote a column for *The Los Angeles Times* about the current administration's plans for using small nuclear weapons to destroy underground bunkers. The administration, Drell said, "were trying to make it sound like low-yield nuclear weapons would be clean, without fallout. It was just pure nonsense, to put it politely."

Drell keeps track of the effects of his Jason reports: "I keep a batting average comparing those they ignored and those they listened to." The ones they ignored, he said, "you raise hell." Raising hell may or may not do any good, as with the ratification of the Comprehensive Test Ban Treaty. "If your expectations are very high, you have to be prepared for disappointment," Drell said. "The ratio of output to input in doing government work is never high." But he doesn't discourage easily: losing a battle, he said, "is not a major disruption in my life. I take great comfort in the fact that for fifty-seven years we've managed severe crises and not used nuclear weapons. I just want to see that we continue to keep our wits about that. And I haven't lost my optimism. I am a great optimist."

CHAPTER NINE

Whither Jason?

"Institutions often outlast the mission they were set up to do."
—Marvin (Murph) Goldberger, interview, 1991

G iven the times, the influence of a Jason report on the president's decision to sign the Comprehensive Test Ban Treaty was a fluke; Sid Drell said it was "simply fortuitous." Beginning with Nixon's second term and continuing with notable interruptions through every administration since, the government has been less and less interested in scientists' advice. In 1995 Congress closed its source of science advice, the Office of Technology Assessment, and never reopened it. By 2004 neither the president's science adviser nor the current incarnation of PSAC, called the President's Council of Advisers on Science and Technology, reported directly—in a networked culture that operates by talking face to face—to the president. Scientists said that government no longer based its decisions—about, say, the wisdom of signing a climate treaty, or the likelihood of an effective missile defense—on independent scientific advice; and advice that was given anyway was ignored. In 2004 the Union of Concerned Scientists charged the government with suppressing, distorting, and undermining science advice; and the Federation of American Scientists published a report on science advice to the government called "Flying Blind." A former

high-ranking defense official said in 2005 that for the past decade
the Pentagon has been calling scientists "tech weenies," as in "these
tech weenies don't understand our complex issues."

The national situation was discouraging, and Jasons found it
so. On the local level, moreover, Jason was having unusually bad
impedance problems with DARPA. And even internally Jason began having problems, not only with its increasing scientific diversity, but also with keeping Jasons around for the whole summer.
Like a newly hatched turtle crawling from its nest to the sea, Jason
had a number of ways it could die.

In July 2001 the newly sworn-in DARPA director, Anthony Tether,
and the previous director, Frank Fernandez, went to the wrap-up
at the end of the summer when Jasons tell their sponsors the outcome of their studies. Fernandez wasn't impressed by the studies'
technical content and didn't think Tether was, either. Within the
next month Tether told Steve Koonin, Jason's chair, that he had
three people he'd like Jason to consider as members.

Previous DARPA directors had also wanted to name Jasons.
Back in 1989 a DARPA director named Craig Fields suggested
some new Jasons who he said would improve Jason's quality. Jason's chair, then Will Happer, told Fields that Jason always wanted
to improve its quality, but that these particular nominees were not
a good match. Happer suspected that Fields's next move might be
to cut Jason's budget. So Happer went to see what he called "a
green-eyeshade person" on a congressional appropriations committee who, when Happer explained his problem, said, "Those
guys. Here, let me write the Jason funding into DARPA's budget for
next year." And "with a stroke of a pen," Happer said, "there we
were as a line item." In political Washington, smallish line items
like Jason can be added or subtracted at will because, as Happer
said, "you know, they're used to pork." When Fields got his budget

with the Jason line in it, he called Happer and said, "All right, Will. It's not much money, and I've got a lot of other problems. Peace." Happer said that he and Fields had been careful to keep the disagreement quiet and mutually unembarrassing: "We had different jobs, and we had to do right by the people we were working with. So that's how we did it."

Except for Fields, however, DARPA directors' suggestions for Jason membership have been posed the way Stephen Lukasik's was: not that Jason accept specific people as members but that Jason beef up its expertise in specific fields. Ed Frieman said that when he was chair, discussions over expertise were a normal part of the Jason-DARPA conversation, and "we didn't take it amiss." Jasons, he said, never needed to "fall on our swords over it."

With Tether's nominees, however, Jason needed to fall on its sword. Jason put the nominees through what Koonin called "the usual screening/vetting process": one was an engineer whose academic credentials didn't rise to stellar; the other two were heads of information technology companies, and neither had doctorates. Koonin sent Tether an e-mail saying Jason didn't think the nominees would work out, and he figured that was the end of it. The following December Tether told Koonin to accept the three nominees or he wouldn't renew Jason's contract. Koonin countered: Jason would take the three on not as members but as consultants on specific projects; Jason produced a list of academically distinguished candidates in similar fields. Tether declined to negotiate. Jason felt blindsided: no one understood why Tether would pick this fight. A war of words ensued, and positions hardened. Frank Fernandez talked to Tether on Jason's behalf, but by that time, Fernandez said, "enough had happened on both sides that the divorce was a done deal." The long, close relationship between Jason and DARPA ended.

Jason had lost not only a sponsor but the channel for all its other sponsors as well. DARPA had supplied about 40 percent—about

$1.5 million—of Jason's budget itself, and if the Department of Energy or the CIA or the navy, for instance, wanted to buy Jason's services, it had done so through DARPA. As of January 2002 Jason had no money. It canceled its winter study. It notified its members that no further work or expenditure was authorized. Then, in "a flurry of phone calls and e-mails up and down the hierarchy in the Pentagon," Koonin said, Jason tried to negotiate a DARPA substitute.

Koonin, Roy Schwitters (Jason's vice chair and a physicist at the University of Texas at Austin), and Robert Henderson (Jason's director at Mitre) spent hours every day on the phone. "We had a big phone list," said Schwitters. "We were scrambling." Others scrambled, too, Schwitters said, "emissaries at very high levels and at intermediate levels," including William Perry, who called the secretary of defense, Donald Rumsfeld. The following spring, the flurry of negotiations converged on Perry's and Herb York's old office, the Office of the Director of Defense Research and Engineering, which sits in the defense hierarchy just above DARPA and just below the Office of the Secretary. Confusingly, somewhere in the process, Rumsfeld gave an interview implying the truth of the otherwise-unsubstantiated rumor that Tether, in taking his stand, had been acting on Rumsfeld's orders.

DARPA and Jason both said that their relationship had been rocky for some time. DARPA had become interested in certain subfields of information technology and biotechnology where the best expertise was no longer in the universities but in industry, so DARPA thought Jason needed more ties to industry. DARPA was also less and less interested in fundamental scientific research. Jason thought that too many of DARPA's studies had become "trivial" or less well defined, and that DARPA didn't seem to care that much anymore.

During the divorce proceedings, someone had told a reporter about the situation; then DARPA had issued a press release accusing

Jason of being Cold War physicists who refused to modernize. So briefly the divorce was news, and Jasons were quoted talking freely. Some press reports found in the divorce the larger issue of the George Bush *fils* administration's attempts to control the independence of scientists. *The Chronicle of Higher Education* cited the Jason story as the latest evidence of "the Defense Department's increasingly hostile attitude toward independent scientific advice." An editorial in *Nature* said that Jasons had refused to accept the "down-grading of their status to that of many other government-appointed yes-persons," and it ended by saying that Jasons provided "the unflinching advice that government officials don't always want to hear."

The DARPA divorce seemed to have had little effect on Jason. After it was all over, Jasons referred to the DARPA divorce as "unpleasantness," "shenanigans," a "dust-up," a "hoo-ha," and "Jason's near-death experience." This is not to say that for Jason the issues that caused the divorce were unimportant. Self-selection is the core of Jason's independence. If DARPA had succeeded in populating Jason with people who could not help but be grateful to their benefactors, then DARPA would have necessarily undermined the objectivity, and therefore the value, of its advisers. And though Jason studies must be determined by what its sponsors need, they must also be determined by what the Jasons are good at. If the kinds of studies that sponsors want Jason to do ever shift too far in the direction of what the Jasons do no better than anyone else, then Jasons will one by one drop out. The politically realistic adjustments Jason had made with other DARPA directors weren't relevant here. If Jason had not fallen on its sword over this issue, it would no longer have been Jason.

For Jason, its near-death experience had been all the more urgent because it coincided with a national emergency. On September 11,

2001, terrorists hijacked four airliners, flew two of them into the
the towers of the World Trade Center in New York City, one into
the Pentagon, and one into a field in Pennsylvania; nearly three
thousand people died. Within a month members of Congress
and the media began getting letters laced with finely ground anthrax; five more people died, twenty-three got sick, and the
postal service was chaotic for months. Jasons, like most scientists, are active e-mailers. Beginning September 11, Bill Dally, a
computer scientist at Stanford who joined Jason in 1993, got so
many e-mails from Jasons that he had to install a filter. The e-
mails first said what everyone else at the time was also saying—
they were "just basically shocked," said Dally. Later e-mails, he
said, were "analyses of why the buildings collapsed, what you
could have done to avoid the collapse, a calculation of the
amount of energy in the fuel of an airliner, analyses of where else
terrorists might strike and what countermeasures the government
should take. It's the way Jasons think about things." Paul
Alivisatos, a chemist from Berkeley who had joined Jason the year
before, said that Jasons already knew the country was vulnerable,
"and in the days afterward I think there was huge anguish that
they hadn't really thought of exactly what happened, that they
should have, that they could have, but that they didn't really—you
know. They wished they had done something." Will Happer said
that the e-mails peaked again after the anthrax letters, and he
agreed that Jasons seemed to feel personally responsible for everything: "I think there were a lot of feelings of guilt within Jason. I
certainly felt very bad about it. Somehow we had failed."

Throughout the DARPA divorce and the slow negotiations with
DDR&E, the biggest worry for most Jasons had been that they
might lose the opportunity to work on systems for detecting terrorists' radiological, chemical, or biological attacks. They'd already
been doing occasional studies on detecting the presence and the

storage locations of conventional and nuclear explosives. What was
new to both Jason and the Defense Department, however, was defending against biological attacks.

In 1986 Henry Kendall—a Jason who had worked on the Vietnam barrier—had made a little joke that was specific to its time:
Jasons were primarily physicists because the weapons problems
were physics problems; among "Defense Department tasks," he
said, "there are not terribly many biology topics." In 1986 this was
like saying the Defense Department didn't need to write terribly
many poems. As of the late 1990s, however, the aphorism about
World War I being the chemists' war and World War II being the
physicists' had a new punch line: World War III will be the biologists' war.

The biologists' war is, of course, biowarfare and bioterrorism,
the weapons being disease-causing microbes. Biological warfare
by one country against another has limitations—clouds of microbes are inefficient killers, hard to target precisely, and hard to
spread effectively—that make it ineffective as a national strategy.
Biological terrorism by and against individuals—the same clouds
released in a subway or on an airliner—is unhampered by those
limitations and therefore is much more effective. In 1999 Jason did
a study for DARPA called "Civilian Biodefense" that predicted how
the latest in genetic engineering might not only change the character of microbes used as weapons but also move the battlefield
from the military to the civilian. The study also said that though
sensors of biological weapons were available, neither the federal
government nor the health care profession was prepared to turn
the sensors into an interconnected system that could detect and respond to an attack. It suggested that the best detection/response
system might be the existing public health system, and it recommended additions to the system. "It was a very good report," said
Frank Fernandez, who was DARPA's director at the time and had

commissioned the report. "It was briefed up the line and other agencies picked up some of the surveillance techniques. The techniques are being pursued."

In preparing for that report, Jason had asked the secretary of the navy to talk to them about biological weapons, which among other things have a restriction that other weapons do not: biological weapons are considered immoral and are banned by international treaty. Jasons asked the secretary how to study a defense against bioweapons without also studying the outlawed offensive bioweapons—for instance, anthrax that aerosolizes more efficiently. The secretary said that was hard but had to be done: testing offensive bioweapons, he said, "is undesirable." Paul Horowitz said that anyway offensive bioweapons were a nonissue, that they are so illegal that no one developing them would sponsor a Jason study. "If anybody's working on that," he said, "they're not going to let a bunch of academic scientists into their knickers."

A study done subsequent to the anthrax attacks, in the summer of 2002, was called "Biodetection Architectures." It began by saying that in five biodefense studies done over the previous five years, Jasons had become increasingly frustrated with what they called the "near-pervasive focus on biodetection gadgetry." What the country still didn't have, they said, was a way to turn the gadgets into a system that could be set up at that moment and in the real world; they suggested ways to create a system that would be "nimble," would be capable of being installed quickly, and would change as the threat changed.

Jason began hiring biologists in the mid-1990s and has since done biological studies not only on biodefense but on everything from the human genome, medical imaging on the battlefield, and cleaning up toxic waste with microorganisms, to the possible links between the navy's sonar exercises and miniepidemics of beached whales.

Jason biologists' collaboration with Jason physicists is a little unnatural. The physicists' training and that of Jason mathematicians, astronomers, geophysicists, oceanographers, and computer scientists have considerable overlap. All physicists are trained in mathematics; all astronomers, geophysicists, and oceanographers are trained in physics. Physicists are now often expert in computer science's hardware and software, making electronics and building searchable databases. Between physicists and biologists, however, lies an ocean of differing training, languages, methods, and subjects. Of all the sciences, Will Happer said, biology is "about the extreme limit that a physicist can interact with easily."

The biggest differences between physics and biology are in the systems studied and in those systems' operating principles. Physics studies the cosmos and assumes it operates according to the universal principles of mathematical logic. Biology studies the illogical, ad hoc, redundant systems of life and assumes they operate according to the principle of natural selection—that is, stay alive long enough to reproduce. A galaxy may be complicated, but it follows clear physical laws; a brain is infinitely more complicated than a galaxy, and its rules are whatever, during its evolution, happens to work. Physicists call biology a soft science; sometimes they call it "squishy."

So when the Department of Energy's Ari Patrinos, who directed both biological and environmental research, began asking Jason to study the government's campaign to unravel the exact order of a human's DNA, Jason once again had to get itself educated. Jasons got hold of a series of taped lectures on biology that they played at lunch. They invited biologist lecturers, including some luminaries in the field of genetics. Jasons also got lessons from

Dr. X on how to sequence DNA. DNA, which commands almost everything that happens in the body, is made of four basic building blocks; sequencing DNA is a way of finding out the order of those building blocks, which in turn is the code for its commands. Dr. X said to the physicists, "Let's do it as a tutorial."

The physicists met at Dr. X's lab and "actually, literally did the sequencing," he said. Dr. X stood at the front of the lab, while the physicists were scattered around at the benches. They pipetted tiny amounts of DNA into a tube, then added enzymes that would help identify the order of DNA's four building blocks. "We were just pipetting things from one little beaker to another," said Freeman Dyson. Then they heated the final mixtures to a precise temperature, put a portion of the mixtures on a gel, ran an electric current across the gel, and mapped the building blocks now separately arrayed in discrete, sharp bands.

Some of the physicists did better than others. "Two got absolute, state-of-the-art-quality sequence data," said Dr. X. "A bunch got fuzzy reads, but you could tell if you knew what the answer was supposed to be. And then there were a few that just sort of got blanks." Dr. X thought those latter few just weren't used to handling biological solutions. "Those that sort of manipulate a solution as a squirt, rather than as a squirt of 2.07 microliters, lose," he said. "And having a tremor doesn't help."

Freeman Dyson did badly. "There was something wrong with the temperature," he said. "I kept it at high temperature for too many seconds." Dyson's lab partner was Roy Schwitters. "My DNA thing was a complete washout," he said. "I was really great at pipetting. I was *good*. But it didn't help me." Claire Max's didn't work either, but what she did wrong, she said, "beats me." Dr. X was nice about it: he told them, "Even for those in the business, it doesn't work all the time." Bill Dally said that only he and Steve Koonin got it completely right. "I think it's largely a test of how well you follow directions," he said. Another Jason, a physicist, said that his was the "prettiest-looking. It looks like a bar code—smears, dim faint things. I was going to take it, but it's Jason property." Koonin solved the Jason-property problem by hanging his bar code on the wall in Jason's offices in La Jolla. "I really got to learn things," Schwitters said. "No, listen, the great thing is, just learning other areas. And we all just love that."

The differences between physicists and biologists, among Jasons, are mitigated by one of Jason's prerequisites for membership: breadth of interest, a taste for the unfamiliar and for crossing scientific boundaries, stated in Jason's charter as the ability to "contribute to unfamiliar practical problems." Jason's current administrator, Robert Henderson, describes Jason to new sponsors as "very smart amateurs." And though this taste for the unfamiliar is especially prevalent among physicists, it occurs plentifully in other scientists. Dr. X says that he's more mathematical than most biologists and has always been interested in physics; besides, he says, "I love interacting with physicists, I really do." Steven Block, a biologist at Stanford who joined in 1996, has a father who's a high-energy physicist, so Block grew up around physics; and his own research is, he says, "a brand of biology heavily influenced by physics. I live with one foot in the physics camp and one foot in the biology camp." A number of Jason physicists also have their feet in biology: for instance, Henry Abarbanel, the physicist who worked on Jason's climate projects, now does research with neuroscientists to figure out how the fourteen neurons in a lobster's stomatogastric ganglion are connected and act together. "It's like being born again," Abarbanel says. "I mean, I am just supercharged. I am the newest member of the graduate program in neuroscience."

None of this means that the collaboration always works. In 2001 Jason did a report for DARPA, called "Biofutures," to consider whether the physicists' trick of coarse-graining—the trick

used in the climate Jasons' Model of the World—would work for
biology. For instance, could the intricate network of chemical signals within a cell be reduced to a model that would allow biologists to predict the cell's behavior? "What we had in mind was not
trying to scare away biologists," says Abarbanel, one of the report's
authors, "but trying to convey to them that there were times when
eliminating the detail, or averaging over it, or lumping it into pa-
rameters, worked very well."

Happer, a physicist who invented a new way to make high-
resolution medical images of the lungs but who was not a coauthor
on "Biofutures," has doubts on behalf of biologists that he illus-
trates with a Jason joke about Murph: Murph was on Jason business at a military base and was treated with exquisite military
protocol. He returned home and told his mother, who happened
to be visiting, "You know, they put me up as a three-star general."
Murph's mother said, "Well, Murph, I think you're a three-star
general and your wife thinks you're a three-star general, but what
did the other three-star generals think?" Murph says that's just "a
take-off on an old Jewish joke." The point is, says Happer, "I worry
a little about our forays into biology."

What the biologists think is that so far, with a few exceptions,
coarse-graining won't work for biology: too little of what needs to
be known is known, and too much of what is known is idiosyncratic, so simplifications or averages are unlikely to be representative. Steven Block, a coauthor on "Biofutures," cites a joke about a
dairy farmer who asked a physicist how to estimate milk production, and the physicist's answer began with "Assume a spherical
cow." The spherical cow fallacy, Block says, "doesn't mean that
physics doesn't apply to biology. It's just that if you make too many
simplifying assumptions, what you're left with may in no way
resemble the biology." Dr. X, also a coauthor, agrees. "Biology, more
so than any of the scientific disciplines, requires just that god-awful

detailed knowledge base," he says. "The physicists will say, 'Well, we can just model all this.' Well, no, you can't actually."

Dr. X says that physicists don't mind being wrong, however, and he mimicked them saying, "Oh. Oh really," and stepping back and coarse-graining again but at the next finer level of detail, and then again at the next. "The hard-core physicist Jasons have come a long way in biology very quickly," he says. So what did he think of "Biofutures"? His pause was unusually long. "Well," he said, and then paused again. "There were parts of that one that were good."

Jason physicists' interest in biology happens to be part of a national trend. Traditionally physicists believe that, since physics is the most fundamental of the sciences, physicists can apply that foundation to problems in other sciences, and in fact physicists have moved successfully, even famously, into biology. Lately, because biological research has been flourishing, physicists have been moving into biology faster. Scientific journals have run a rash of articles on physicists taking their skills into biology, attended by the same problems that Jasons have had: an editorial in *Nature* was titled "Pursuing Arrogant Simplicities"; an article in *Physics Today* was called "Harness the Hubris."

In fact, the national trend of physicists moving into biology is itself part of another national trend. In spite of an opposing tendency toward increasing specialization in a single field, much new science is now being done at the junctions between different fields. National funding agencies, like the National Science Foundation, and academic centers all over the country have mission statements that feature the word *interdisciplinary*. Young scientists, including young Jasons, who do tomorrow's science are increasingly interdisciplinary. Prof. Y, who joined Jason in 1993, trained as a chemist but works as a condensed-matter physicist; Paul Alivisatos trained as a physical chemist and says he speaks the languages of both physics and chemistry and adds, "I have great desire to know and learn biology as well. I try to be as unclassifiable as I possibly can. For entertainment's sake." Dr. X says that his interests have never stayed within one discipline: "So the biologists think that I'm very chemical in my research, and the chemists think that I'm very biological in my research, and that's just fine with me. I've never tried to worry about where the line is."

If science has changed, and physics has changed, and the character of warfare has changed, and Jason's sponsors have changed, then Jason had better be changing, too. And once again it has been: compared to the 1960s Jason, the change is startling. Over a third of Jasons are not physicists. Only around 20 percent of their studies are strictly on physics; about 15 percent are on biology or are heavily biological; about 50 percent are on computer science or signal processing; and the remaining 15 percent are multidisciplinary mergers of all of the above plus chemistry. Every Jason said that Jason's biggest change was its new diversity, none of them regretted it, and given the increasingly dicey probability of staying impedance-matched to the country's problems, none thought they had a choice.

Every time Jason took on a new problem for which it lacked expertise, it was forced to broaden its self-concept. Paul Horowitz mimics Jason considering its options: "We can either say, 'Well, we're not going to be able to help you because we don't know anything about how you deal with terabytes of data.' Or we can fake it and say, 'We think we're computer scientists because we all have computers on our desks.' Or we can get with the program and say, 'We have some people who do this for a living, and we'll give you a knowledgeable and up-to-date answer.'" Jason chose to get with the program. "That's the reason we changed," Horowitz says, "we didn't do it just for the fun of it."

Jason's scientific diversity, which began with the Jason Navy and grew through the Jason climate studies, now with the biologist Jasons seems permanent. One side effect is that new members, who are coming from a wider pool, haven't necessarily had previous long-standing academic ties with other Jasons. Jason no longer confines itself to physicists' old boys' networks; its members are now not only not physicists, they're not boys and they're not in the network. Paul Alivisatos and William Dally came to Jason's attention only because they were members of IDA's Defense Science Studies Group, a rotating group of young academics that advises the Defense Department and that is advised in turn by several older Jasons, including Callan, Happer, and Koonin. Peter Weinberger, who joined in 1990, knew no Jasons until he met Bill Press on a government committee. Dr. X knew the work of one Jason and happened to socialize with several others. Prof. Y knew no one; David Nelson, a physicist at Harvard who'd been a Jason since 1985, heard her give a talk and thought she'd make a good Jason. And as with any anastomosing process, the Jasons who invite new Jasons are themselves less related than the generation who had invited the inviters: each new generation just gets farther apart.

Nevertheless the Jasons neither say nor give the impression that they feel less communal. Despite their scientific diversity and because of their shared taste for the unfamiliar, they work together as effectively and happily as ever. "I don't perceive a problem," says Happer. "Some of the new Jasons are really great," says Peter Weinberger. "So if you bring people in and they interact, then I think Jason works fine." *Interact*, though a perfectly good common-usage word, is also physics jargon for two particles with certain characters and trajectories that converge until they feel each other's fields of force, then either collide or veer off; in either case, their characters and trajectories have changed.

The change that might endanger Jason, that all Jasons talked about most, was not scientific or military or political but sociological.

After Peter Weinberger said he thought that with new Jasons inter-acting, Jason worked just fine, he added: "Now before I got there, I am told, and in an era that's past, the families would show up for the summer. And that's gotten a lot harder." Every Jason says the same thing, in the same order: in the olden days Jasons were all men; they put their jobs on hold, packed up their wives and children, and de-voted six whole weeks to working with other Jasons; now things are different. In the 1960s, says Steven Block, "Jasons had wives who didn't have jobs, and they could go off to someplace beautiful like La Jolla. The Jason stipend was large enough that they could rent a cot-tage by the beach, and the kids would spend the summers gamboling in the waves, and wives would take care of the kids. And the husbands would join the families in the evenings for swims. That was a differ-ent world."

The first and biggest difference lies in the portability of fami-lies. Physicists tend to be familial: an anthropologist/historian who studied physicists said that their professional culture includes their families, that "a successful physicist is a married physicist." So from the beginning Jasons acted as though their families were part of the deal. They picked meeting places near beaches, rented houses, gave dinners, had picnics, played baseball, and celebrated holidays. They'd have a party when the summer started and a party when the summer was over and parties in between. Walter and Ju-dith Munk usually had Fourth of July parties and put on plays in a backyard that is an amphitheater sculpted out of the hill and over-looking the ocean; at night they'd all watch the fireworks from there. On weekends the families would go hiking in the desert, climbing in the mountains. The wives became friends, Mildred Goldberger said, and the children grew up like cousins.

"My kids all have their Jason kid-friends of comparable ages that they've kind of grown up with over the years," says Bill Dally; several Jasons talked about their own children and Jason coevals conversing throughout the year with e-mails and instant messages. Dally's wife, Sharon, has come to La Jolla most summers: "So she'd say, 'We're having tea at Pannikins today, and then we're going down to Balboa Park.' And I'd say, 'I'm going to sit in a room with no windows all day and work on things I can't tell you about.'" Will Happer's wife, Barbara, is a school nurse with summers off and has joined a group of local people who rescue stray dogs: "To tell the truth, and I don't think I'm the only one, the fact that our wives like to go to La Jolla is a big part of being a Jason," Happer said. "Like I couldn't retire early very easily because Barbara would be upset."

During the summers in La Jolla, one or another of the Jasons or their spouses organizes informal evening lectures by guest speakers called family talks. After the talks are snacks and drinks on the lawn. "The kids below a certain age don't sit through the talks but run around on the grass outside," says Press. "We usually pay some of the teenagers to watch the kids. It is a family thing." To the outsider, Jason sounds a tiny bit smug, the way other people's happy families often do.

But the millennium turned, and now three or four Jasons are women whose husbands can't leave their jobs. Prof. Y's husband is an astrophysicist on the East Coast; when Prof. Y goes to La Jolla, she brings her mother along to baby-sit. Claire Max's husband is an astrophysicist at Berkeley and comes to La Jolla for only a few days here and there. "My son came down with me every year," Max says, "and he still keeps in touch with the kids that he knew. I feel very warm toward not only the Jasons but their spouses and their children. It's hard for me to separate the intellectual community from the family community. But it's not as easy to do now as it was twenty years ago."

Jasons' wives can't leave their jobs, either. Working wives are no longer the exception but the rule. "My wife runs a twenty-person firm," says Rich Muller, who lives in Berkeley. "She can't take the summer off just to keep me company. I commute to Jason, I come home every weekend." Roy Schwitters's wife comes to La Jolla "a little bit of the time," he says. "In some sense, I get back into graduate student mode. I work and crash on the weekends. It's very intense. But in this perverse way I enjoy that." Steve Block at Stanford doesn't enjoy it as much and says—as do other Jasons—that the Jason summers put his marriage under pressure. "It's been a huge sacrifice for my life. Your kids can't come along, your wife can't come along, you're reduced to living in these residence hotels. Oh God, if I see the inside of those walls one more time. So I've been cutting down the amount of time I spend there, and I go home occasionally on weekends."

Jasons' jobs have also changed. The first Jasons were theorists whose tools were relatively portable—pencil and paper, calculators, their heads—and who therefore often spent summers at workshops in places like Aspen or Sicily. Happer repeats an experimentalists' joke: "the leisure of the theoried class." Jasons increasingly are experimentalists, however, who are tied to their labs. They work year round running expensive, delicate experiments in labs, deciding the experiment's next step, supervising the graduate students who do the hands-on work. "There's nothing like face-to-face interaction, sitting down and going over the data. It's not good to be away," says Prof. Y. So when Prof. Y is in La Jolla, in addition to her Jason work, she spends hours in e-mail contact with her group. Biologists, engineers, and computer scientists operate the way experimental physicists do. "I need to see data almost every day in this lab," says Dr. X, the chemical biologist. "You can't run it by e-mail and pen and paper." Dally, a computer scientist, says that when he finishes his eight-to-ten-hour Jason day, he puts in another three hours on his research.

Added to immovable spouses and inflexible jobs are a few

lesser differences. La Jolla has gotten more crowded and more expensive—"too chichi," says Block—so the Jason housing allowance now covers houses not on the beach but in the hot, dry, overdeveloped interior; driving anywhere means crawling through an infinite traffic jam. The pace and demands of academic jobs for theorists and experimentalists alike have risen; scientists travel more often to more conferences; grants are harder to get; scientists have more competition. "Everyone's running around all the time," says Paul Alivisatos. Computer scientists—who are hard for Jason to get in the first place—and physicists occasionally start their own companies, so Jason summers have what Dally calls a "high opportunity cost": missed opportunities for the company. Dally helped start at least three companies; a new Jason whom Dally invited has started several and, Dally says, "he's still got his hands in some of them. I doubt we've gotten two weeks out of him yet, during a summer study."

In short, the various difficulties with spouses and jobs all come down to the Jasons spending less time at Jason. Prof. Y, who is on the steering committee, estimates that only a half to a third of Jasons stay for the whole six weeks. Bob Henderson says he doesn't have any hard numbers, but attendance decreased throughout the 1990s. The local Jasons from San Diego, Henderson says, "come in one day and out the next." The semilocals from Stanford, Caltech, and Berkeley are commuter Jasons; they come for a slightly abbreviated week, in Monday afternoon and out Friday morning. From the truly out-of-town Jasons, Henderson says, "we don't get six weeks as often as we used to. We might get three weeks or four weeks." Some Jasons show up even less: Dally, from Stanford, calls them "cameo Jasons," Jasons who "show up, do a little bit, and then leave," and says he's one himself.

Dally says that though being a cameo Jason is good for him personally—"I'll swoop in, and somebody will say, 'Here's what we need you to do for our study,' and it'll be fun"—it's hard on Jason. Have the cameo and commuter Jasons affected the quality of the studies? "We're not subcritical," says Henderson. "We still do good things, useful things. But could they be better? I think some of them might be better if we had more people and more time." Dally says that if everyone did what he himself does, "Jason wouldn't work."

Jason's survival depends first on whether its sponsors can find the same quality advice elsewhere. What sponsors can find nowhere else is a high-quality, self-selected, professionally and politically independent group that is possessed of the sponsor's corporate memory, that has members with high clearances used to working together, and that does the work itself and spends weeks doing it. Subtract a few of those characteristics, and sponsors now have plenty of alternatives in the Defense Science Board, the advisory boards of the military services, the National Academy of Sciences, and the Mitres, RANDs, and IDAs. DARPA has two Jason-like groups of advisers that include stellar academic researchers; one focuses on information technology; the other, nicknamed the Masons, focuses on materials science.

But so far none of the alternatives offer everything Jason does. The members of the Defense Science Board are selected by the Defense Department and often come from the defense industry; their reports are done by staff. The advisory boards of the services meet for a few days at a time. The National Academy of Sciences prefers unclassified studies and doesn't do its own research. The Mitres, RANDs, and IDAs are large, permanent institutions tied to specific sponsors; their reports can take years. DARPA's two advisory groups are bound more tightly to DARPA than Jason was, and they meet for shorter periods. Most of the other advisers don't invent: their members come to the meeting, Curt Callan says, speaking a little loosely, "and disgorge two or three pages of what they already know. And some editor puts the thing together and writes an executive summary, and everybody approves of the executive summary." Happer says he knows of no other organization whose members "dig in and actually solve differential equations," working together for weeks.

In short, if Jason's survival depends on its external usefulness, then its survival and its external usefulness both depend on its internal cohesion. And its cohesion—given new national problems, immovable families, inflexible jobs, and commuter/cameo Jasons—is what's at risk. Enough lost cohesion, and sooner or later Jason will lose not only its sponsors but its own sense of community.

The steering committee talks about this problem regularly. "Absolutely," says Schwitters. "We talk about it, and then we talk about it. When we talk about it, we say, 'Well, what are we going to do?'" These meetings come under the same heading as the Whither Jason meetings that Jason has held periodically. Jasons debated their position about Vietnam at Eglin Air Force Base in a Whither Jason meeting; Jason's move into climate science was the result of a Whither Jason meeting; the group held Whither Jason meetings when the Cold War ended. These meetings have usually been about balancing the sponsors' needs against Jason's abilities and preferences: what studies need to be done, what's already being done, what's being done poorly, what ruts Jason is getting into, what risks becoming a job shop, what's new and interesting. The current Whither Jason meetings, however, are about sociology, and so far nobody's had any bright ideas. "I think Jason, if they could find a simple answer to it, they would do it," says Alivisatos. "But you know, a macrotrend within society, how does Jason solve that? It's not obvious, okay?"

They do have ideas, of course, though Dally says that every time somebody suggests a new idea, "it tends to get shot down," and if the person who suggested it doesn't shoot it down, another Jason does. Block thinks that in years to come, "Jason may have to pack it in and figure out another way to get its work done besides these intense, long meetings in La Jolla." Ed Frieman says Jason should try the technology that scientists already use to work with colleagues at a distance: "Many people these days hold teleconferences—one can actually do that in a classified way—or they web-ify everything," he says. "All sorts of technology now exists." Prof. Y says teleconferences don't allow people to work the way Jason needs them to work: "If you just take an hour away from your work and go sit in a teleconference and then go back and work on what you were working on before, it's not going to go." Schwitters is also dismissive: "We have some of those tools, we can do them. But you know very well, the presence is still important, the human give-and-take in these situations."

Some have suggested distributing the studies so that only a few Jasons would need to show up, then have them stay only long enough to do the study. Block says that distributing the studies would reduce the chances for cross-fertilization among Jasons working on different studies; Prof. Y says, "you really would lose the kind of social cohesiveness of being together for some period of time"; Koonin says he prefers "the more contemplative, sometimes serendipitous mode." Considering that each Jason generally works on four to six studies a summer, distributing the studies also sounds logistically unlikely. Nor can they save time by focusing the questions tightly from the beginning. "You don't just say, 'Okay, I have a question and now I'm going to answer it,'" says Prof Y. "Very often, figuring out what questions to ask is the single most important thing you do."

With no good solutions in hand, Alivisatos says, "it may just be that they'll end up finding the people who can make it work."

Prof. Y says, "Either people will adapt to this way of life or they won't." To be sure, some seventeen new Jasons, whom Schwitters says are "terrific," have joined in the last eight years. "Essentially at the end of the day we look at each other," says Schwitters, "great people coming in and we're able to answer the mail. So we'll keep going. We don't have a better idea."

If the question is whether Jason will stay together, the answer probably lies in why it's stayed together this long. As of January 1, 2005, Jason was forty-five years old. Murph hasn't actively worked on studies since 1967, but he's been coming to meetings most of the forty-five years. Walter Munk has missed no summers since he joined in 1961; Ken Case, who joined the same year, missed only two or three; Sid Drell, who joined in 1960, missed a few in the middle of his career but for the last fifteen years has missed none. Freeman Dyson, who began working regularly in 1965, thinks he's missed five, but none for the past twenty years. Richard Garwin attended regularly, though for only a few weeks each summer, from 1966 until 1993, and every summer since then he has spent all six weeks.

To stay in Jason, they've had to give up a certain amount. Not only are they under pressure from their families and employers, but summers are their time for research: running experiments, calculating, writing grants, writing papers, thinking. Some—not all—Jasons think that Jason has hurt their careers. Bill Dally says he thinks his reputation in his own field would have been greater without Jason, though "it's not like my reputation is sullied and Stanford is going to write me a letter tomorrow dismissing me from my tenured position. But there are certain prizes that I aspire to, that I've probably reduced my chances for getting. Because I worked on a Jason study that nobody will ever know about." Will Happer says that his colleagues "who kept their nose to the grindstone, managed big armies of postdocs and graduate students, I think got more recognition. It's probably hurt my scientific career a bit."

So given the costs, what keeps a group going for forty-five years? The first answer is surprisingly simple: they like the work. They enjoy detecting lemons, they think inventing is fun, they love learning new subjects. They like what Dyson calls "real, honest, detailed, technical work" and what could be thought of as getting their hands dirty.

David Nelson's explanation of this is the neatest: "The problems that I worked on for Jason were seductive. It's like saying to yourself, okay, you're a physicist. You teach all these basic concepts. Now let's see you *do* something with it." "Doing something" means thinking "more like an engineer," the way Archimedes and Leonardo da Vinci did, "not that I'm in their league, but it's nice to have that different take on science and the world. And why can't you do the whole thing? Why can't you be a scientist, have a nice idea, work out its consequences, and see it through? Why don't I actually try to do something with it that would help the country and be useful?" Nelson also understands the moral choice implicit in thinking like an engineer, because he adds that he has never worked on anything that seemed pernicious: "I worked on fairly innocuous and straightforward problems. In fact, all the Jason projects I knew about during my time—mid-eighties onward— seemed sensible and important to me."

Another answer to why Jason has lasted is, the Jasons like each other. "What keeps the Jasons going back to Jason are the other Jasons," Block says. "They're remarkable people," says Walter Munk, "and having them as friends is a real privilege." All the Jasons say this, old and young. "I have to say one other thing about being in Jason," says Prof. Y, "and that's the opportunity to be around people of the quality of Sid. That man is so remarkable. And many of the older Jasons, like Will Happer, are just remarkable people. And the chance to interact with people like that—what a life experience." I asked Weinberger whether he liked working with certain Jasons. "I like most Jasons," he said. "I think that's why I stay. It's a very interesting group of people to be around." The words Jasons use most about other Jasons are *remarkable* and *refreshing*.

They also work well together. "It is delightful to work on a project where the aim is to get a job done rather than to get the credit for it," says Dyson. "Most of us function best when working in a group, especially when the group is bonded with friendships. Most people are happier cooperating than competing." Horowitz says, "Ego's not in this thing. And the better we can do as a group, the happier we are." Prof. Y says that the pleasure is not that Jasons have no egos, but that they have no problems with their egos: "None of these people are insecure. It's nice just to be among people who aren't threatened by other people who are doing well."

The third answer is, they are patriotic. *Patriotism* is a word so ubiquitous as to be meaningless. It is such a cliché that it's understood as an obvious cover for less high-minded motives, like wanting money, or liking to know secrets, or having access to the powerful, or being one of the intellectual elite. Jasons have all those motives and don't mind admitting it; if they denied it, no one would believe them anyway. But just because patriotism is a cliché and only part of the truth doesn't mean it's not still true.

Ed Frieman was born and raised in Hell's Kitchen, a part of New York City as poor and crime-ridden as the name sounds. His friends from those days are mostly dead or in jail, he says, "and I don't know how I ever got outa there." Around 1943, when he was seventeen, he joined the navy and was assigned to the V-12 program, the navy's equivalent of the army's Special Engineering Detachment, which "jammed us through Columbia in two years" and then sent him off to the war in the Pacific. After the war Frieman had to stay in the reserves and was assigned to the Atomic Energy Commission and told to report to a job at Princeton, which turned out to be John Wheeler's Project Matterhorn. One part of Project Matterhorn evolved into the Princeton Plasma Physics Laboratory, of which Frieman ended up as deputy director. He was made a professor of astrophysical sciences at Princeton, then director of DoE's Office of Energy Research. When he says, "This country was good to me," you can believe he means it. "This country gave me an education," he says. "It enabled me to have a decent life. And this is a payback. There's an element of that, I think, in many of the Jasons."

Bill Press has had the opposite life, born with the academic equivalent of a silver spoon. Press's father was a geophysicist at both Caltech and MIT, science adviser to President Jimmy Carter, and the president of the National Academy of Sciences. Press was raised around Caltech, graduated in 1969 from Harvard, did his graduate work at Caltech, spent two years on the physics faculty at Princeton, then was hired back at Harvard, where he was the youngest tenured professor. "For me, having been in college and graduate school in the Vietnam years gives a kind of poignancy to 'national service,' " Press says. "I had acquaintances killed in Vietnam, and I was sitting fat and happy doing science at Caltech. And maybe I owe some equivalent national service. And I'm not in any way trying to equate life in La Jolla with being a foot soldier in Vietnam."

Paul Horowitz's career has been the prototype of the successful academic scientist's. He was a "nerdy kind of kid," he says, who liked optics, ham radio, rockets, and "the wet chemistry part of photography." His undergraduate and doctoral degrees were from Harvard, and he's tenured there; he wrote the standard textbook on electronics. "I think that we basically understand that we're really doing this for someone else," he says. "And the someone else is the government, and we believe that it's our government and that we think our government will always be better if the technical advice it gets is impartial, disinterested, accurate." Scientists can give that kind of advice, he says, because their research, which is publicly funded, has given them a valuable perspective. "The perspective's available, it's in this country, the government's made it happen by supporting research, and they damn well deserve to get the benefits of it back."

Peter Weinberger's career was not a typical academic's. He trained to be an academic mathematician but didn't get tenure, which was not, he says, "a big injustice." He then went into industry, writing software first for Bell Labs, then for a Wall Street hedge fund, then for Google. "I think the cohesion in Jason comes because people want to help the country," he says. "I mean, nobody makes a big deal of it. But I don't think you get involved in Jason unless you have some feeling that there's a chance to help the country. Why would you bother?" Weinberger proposed a thought-experiment: what other setup could attract and keep people like the Jasons? Maybe some industry like General Motors could pay them so much that they'd join for a while, but he didn't think money could keep them interested. "I can't imagine anywhere else. When you get down to it for many of them, I bet it's patriotism. Because what else could it be?"

All Jasons say this or something like it; and maybe it's the answer to a question that I asked them repeatedly and that, after enough nonanswers, I finally stopped asking: if Jasons remain members because they like being useful, why has Jason never put much official effort into finding out whether its studies are ever used?

The nonanswers are usually a list of the exigencies that affect a study's impact—for example, a sponsor's changing priorities or a prior commitment to a program or internal politics; Weinberger says that assessing impact is a "very noisy signal," meaning he can't pick out a signal from the background noise. One reason they don't formally assess impact is that the different Jason clusters already know—that is, they can tell by the new problems a sponsor brings whether their old solutions were any good. Another reason is that while sponsors will sometimes cite a Jason study as backing for a decision they've made, they rarely announce the considerations that went into making the decision in the first place. So if a Jason study affected a sponsor's decision, Jasons might never know it.

I think another reason is that some studies—like adaptive optics, or Bassoon/Sanguine, or ocean acoustic tomography, or Patrinos's ARM measurements of climate change, or the Comprehensive Test Ban Treaty—have enough visible impact to keep Jasons hoping they're doing the country some good, and they do the next study on the basis of that hope. Weinberger himself didn't join Jason until Bill Press had asked him twice: "the first-level problem with Jason was bombs, right? Which are morally dubious." The second time Press asked, Weinberger agreed to try it, then decided to stay. "There was always the possibility of doing some good. One is not exactly in a position of power to do good. But at least you can do your best to give good advice and hope somebody will take it."

The reason Happer has stayed in Jason for nearly thirty years, he says, is "because I thought it was a service to the country." Happer's colleague in Princeton's physics department, Curt Callan, calls Happer "a natural," meaning that Happer is a public servant, Callan says, "because it's in his nature."

Will Happer is skinny, smiles all over his face, and gives the impression of fineness—fine hair, skin, and wrinkles—and youth. He's never quite still; he scrooches around in his chair, slouches, winds his legs around the chair legs, and wraps his arms around his neck. He says things he probably shouldn't, and if you use those things to hurt him or someone else, then that's the kind of person you are. He himself seems utterly unimpeachable.

Happer had an unusual upbringing. He was born in India to two doctors; his mother was from the North Carolina Piedmont, while his father was a Scot in the British Army. When World War II broke out, Happer's father had to serve, so Happer and his mother left India and went to live with her brother in Oak Ridge, Tennessee. At the time Oak Ridge was one site of the Manhattan Project; Happer's mother was the first doctor at Oak Ridge, treating cases of radiation sickness; her brother and Happer's uncle, Karl Morgan, was a physicist there. Happer remembers the mud, the boardwalks, and the guards with machine guns. He was four years old and surrounded by people who knew about how to put pipes together to hold uranium hexafluoride and how to blow glass, and an uncle who explained to him about splitting atoms. "I took it in, I listened carefully," he said. "It was my impression that physics was probably something interesting to do."

After the war the family returned to India for a few years; then when India became independent, they lost their savings and moved along with the rest of the British professionals back home where there were not enough jobs for them, and none for Happer's father. Eventually the family—now with two more children—packed up again and moved to his mother's old home, North Carolina, where his father had to get recredentialed and finally, in his mid-fifties, found a practice. "It was sort of the classic impoverished, insecure childhood that so many people have," Happer said. "Makes you want to avoid it if you can in your own life. Maybe slightly paranoid, you know, about the cruelty of fate." With no financial help from his family, Happer got himself through college, then through graduate school at Princeton, and finally to a job at Columbia, where he was barricaded in his lab by the Vietnam protesters and joined Jason. Eventually he took a job back at Princeton and stayed.

What he liked about Jason, Happer says, was how different its studies were from his own increasingly specialized research. Jasons impressed him with their breadth of knowledge, knowledge that the average physicist didn't have: "A number of them, for example, were pilots, so they knew from the seat of their pants about lift and drag and trailing vortices. I liked to go stand by the coffee and just listen to them talk. There's a beautiful law about the lift of a wing that's like a sonnet almost, it's so pretty. It's the most fundamental conservation laws of physics—when a fluid goes faster, its pressure goes down, the famous Bernoulli effect. And that was a surprising thing to me, that these fundamental laws were still at the center of these complicated things." He goes on to explain how those laws operate even when the complicated flow over a wing goes wrong, and the wing stalls: "I learned that when a wing stalls, it's not like when your car stalls. It means that the wing stops lifting, and the plane falls to the ground." These laws of physics "suddenly made something that was so complicated and intractable and inscrutable," Happer says, "crystal clear and self-evident."

Scientists want to know what things are made of and how things work; and when scientists figure that out, they see that under a messy and sometimes lethal world is order and beauty. Jason helped Happer learn that, he says: "Jason was more like what I thought physics might be, you know, when I was a little boy in Oak Ridge."

One of my interviews with Happer was in October 2001; Happer was then slightly obsessed with the September 11 attacks. No matter what the question, he kept coming back to who the attackers were, why the attacks occurred, how the country had failed to protect itself, and how Jason had failed. "I guess I'm more pessimistic that there will ever be a time of universal peace, when there are never any threats," he said. "I read Russian poetry, and Alexander Blok has a great poem, 'On the Field of Kulikovo.' It's about a soldier in the army, and he's thinking about his wife and he's thinking about the upcoming battle. One of the lines goes, 'Peace is only an illusion; there's always a threat, there always will be.' "

Happer didn't think he was quoting correctly and later looked up the poem and translated it from the Russian himself. Professional translations of the poem are florid and bombastic; in Happer's translation, the poem is frightening.

And the battle is eternal! We only dream of peace through
the blood and dust. The steppe-mare flies on and on, and tramples
    the tall grass.
And there is no end. The miles and cliffs flash past. Stop!
On and on march the frightened clouds, the sunset is bloody!
The sunset is bloody! And blood streams from my heart!
Cry out, cry out, my heart.
There is no peace! The steppe-mare dashes away at a gallop!

"When I read that," Happer said, "I think a lot about what my life has been like. And what it's likely to be like for our children." He sat quietly for a while. "Being a Jason, I think about all sorts of unthinkable things. And I don't know what the answer is. Except, as they say, eternal vigilance."

Prof. Y got interested in science in high school; she liked its orderliness and, "you know, the challenge." Her doctorate is in chemistry; she has the title of Distinguished University Professor in her university's department of physics. Statistically, she's a triple outlier: she's not only one of five percent of American physicists who trained as chemists, she's also one of five percent of full professors of physics who are women, and one of the six percent of Jasons who are women. "In Jason it's just a matter of how effective you are," she says. "It doesn't matter if you're a woman. Nobody cares."

She wants to be called Prof. Y because she has worked on Jason's classified studies and, she says, "I know that some of our sponsors would be put off to see my name publicly identified." She knew that this book already had a Dr. X and at home one night was wondering out loud whether she should call herself Dr. XX and if so, then maybe Dr. X could be called Dr. XY; and her son said, "Mom, you are such a geek." She's tall, has silver-blond hair pulled back, uses no makeup, and wears uncluttered clothes. She might look plain if she didn't move so gracefully and if she didn't sit so quietly, self-possessed and intense and honest to the bone.

She joined Jason out of both curiosity and competitiveness, she says, and stayed because "there's a lot of very interesting kinds of things to think about." Her first summer "was mind-boggling. First of all, the physical plant there is very unimpressive. You go inside, you're thinking, 'This is Jason, Jason, Jason,' and your office consists of an empty room with one of those fold-out tables and a chair and a bookcase and a telephone. And you think, 'Oh. Right.' And all the Jasons are sitting around and they're talking in this kind of shorthand about things that they've known for twenty years, and you have no idea what they're talking about. It's not physics, and they're using acronyms, and I couldn't imagine I could possibly make a single useful contribution to anything I was hearing. I found myself feeling very shocky."

The next summer she worked on a study for which she had some background and wound up feeling better than she had the first summer. Her fourth summer she led a study whose topic she proposed, and "it was a riot," she says. "People got interested in it. And at a certain point I realized that even though all these very smart people all around me are infinitely giants in the field, there's stuff I know that they don't know, and it's valuable to them that I'm here."

Prof. Y thinks she joined Jason at a nice time, in 1993, just when the Cold War ended, and Jasons "were going into arms control in a heavy-duty way." She worked on Jason's stockpile stewardship studies and thought the group helped the country reach "a rational state of what was going to happen with our nuclear plants," so that by being in Jason, "I could contribute something good." Anything that helps our nuclear plants avoid irrationality is good.

"I'll tell you something non-Jason," she adds, "to give you a perspective on my orientation." Just before she joined Jason, her university's student newspaper ran an article saying that faculty members who accepted grants from the Defense Department were immoral. The issue is, of course, perennial; it was raised by the chemists' and physicists' wars and again by the Vietnam protesters. To further discuss the issue, the students and faculty held a forum. When Prof. Y walked into the forum, a researcher was taking the scientists' usual position, explaining that he could accept defense money without moral qualms because he worked only on basic, or pure, research. Prof. Y's turn to speak came next. "Frankly," she told them, "this business of saying I can take DoD money because what I'm doing isn't important to DoD, I think is hogwash. It's hypocrisy." Moreover, she told them, "I am a scientist who accepts a lot of DoD money. But before I'm a scientist, I'm a U.S. citizen. And if I'm a U.S. citizen and I have a problem with what the DoD is doing, then it is my responsibility to protest that." In fact, she told them, "as a scientist and as a citizen, I think that DoD is an important part of our society. And because of that I can accept DoD funding without any challenges to my conscience."

Like it or not, the world holds people who are prepared to use weapons. Because a defense department is therefore necessary, advising it is morally defensible. For Prof. Y, advising on stockpile stewardship meant studying "weapons that, if they are used, will have horrendous consequences for real human beings," she says. "That's a tough thing to face." But the alternative of claiming moral superiority and refusing to face the reality of those weapons is an abdication of responsibility. "Nobody's innocent," she says. "Everybody's responsible."

Still, "innocence," if it means avoiding applied defense problems for pure research, is scientists' prime motive. "Will Happer says science is beautiful and pure," she said, "and I think that feeling is why a lot of us went into science." Prof. Y thinks that in an earlier world the people who now become scientists might have been monastics: both professions share a faith, she said, that "there's truth and there's right and we can find it." Scientists say this about finding truth particularly in arguments with humanists who think truth is unknowable. Scientists get away with it because they limit "truth" to that which is knowable and quantifiable, that for which they can find physical evidence, and that for which, when they've found and measured it, they can agree on its interpretation. The problem is, Prof. Y adds, "holding on to that purity and walking with society is difficult."

When scientists walk with society, says Prof. Y, science risks being politicized, its conclusions overstated, overinterpreted, or unwarrantably dismissed in service of political aims. Scientists who can't navigate that politicization won't have their advice heard, she says: "ignoring that society is going to politicize the science is dangerous." But "succumbing to the politicization is also dangerous": too many politically advisable adjustments, and society will forgo the one thing scientists are good for, their ability to build evidence into knowable truth.

For Prof. Y, the way to navigate the political dangers is to stick to who she is and what she does best. "The thing I feel I can do in Jason is, look for the evidence and present it clearly," she says. "Tell the truth. It sounds trite, but it can be a wonderful focus for sorting out an issue. If I tell the truth, even about frightening weapons systems, that's got to be the right thing to do. That can't cause evil."

Prof. Y thinks an interesting book might be written about Jason's history, beginning with the early days of missile defense and nuclear tests, showing how Jasons define their moral balances about working on things "as horrific as nuclear weapons." She thinks such a book might take Jason through the Cold War, through the test bans, and then through the bigger issue of the country's cycles in how it sees science and science's value, and how Jason managed to ride it all through.

Say the Jasons decide to ride it through: Jason's survival still depends on whether the U.S. government wants to keep listening. Granted, the government has no other group that matches Jason for scientific brilliance, breadth, and independence. But the rising sea of science advisers that Steve Lukasik talked about, in the defense industry or in house in the Defense Department, has continued to rise. And the government has increasingly seemed less interested in hearing the advice of scientists who are unaffiliated.

Jason happens to be unique to the United States. John Wheeler and Gordon MacDonald tried for years to help set up a British Jason, but nothing came of it: the differences in the relationships between science and government in Europe generally and in the United States, MacDonald said, are historical, institutional, and cultural—in other words, irrevocable. Pure scientists in the United States are comparatively tolerant of applied science. Pure scientists in Europe are more likely to see applied science as second rate, and as a result, superb technical institutes like Caltech and MIT have no international peers. Pure scientists in Europe rarely advise the government; those scientists who do are usually employed by it.

If the U.S. government stops listening to its independent scientists on political decisions about scientific issues—like climate change, missile defense, or nuclear weapons—the advice that gets heard will come from scientists who know what their employers want to hear. The risk is that the government will be left to make its decisions based more on the political realities than on the physical ones. So it refuses to sign international climate treaties regardless of whether the earth is actually warming. It spends billions on missile defense systems regardless of whether they'll ever work. It advocates the use of tactical nuclear weapons regardless of whether the weapons fit the circumstances and how far the fallout will spread.

The only place the government can get science advice that is reliably apolitical, say all the insiders—including York, Ruina, Flax, Lukasik, Patrinos, Perry, Fernandez, and others who don't wish to be named—is from something like Jason. Not that the insiders universally love Jason: they say that Jasons are often arrogant and naïve; that Jasons admire complexity in a problem and as a result often don't provide what government needs, which is a fast, either/or solution; and that anyway, as Perry says, outsiders rarely turn the tide on a decision. But they also say they can trust Jason to give them answers that are objective and not politically expedient. And when a Jason answer is negative, especially if Congress or the news media hear about it, only strong personalities or sufficient authority can "drive past it," one insider says. "Once they make a recommendation, you've got to get your act together."

William Perry is a member in good standing of what he calls the "interlocking directorate of advisers" and has worked in the Department of Defense, the defense industry, and academia, all three. He has a doctorate in mathematics; he was even a Jason briefly, in the early 1980s, after he stopped being DDR&E: "I thought it was a neat idea. I went to the first summer meeting, but then I didn't think I could commit enough time to make a practical contribution." Perry says that the government doesn't always want to hear from its outsider scientists, especially when an issue has, for political reasons, already been decided. "But if you're still making up your mind, you want the independent group," says Perry. "Jason's a tremendous group, just as smart as you'd like them to be. It would have been an enormous mistake not to listen to them."

On a matter of policy or morality, I would trust Jasons no more or less than I'd trust any educated and thoughtful person. On a matter of science, however, I'd trust them implicitly. I'd trust them even when the science implied policies distasteful to the most fervently pro–climate treaty, anti–missile defense, pro–test ban Jason. If the Jasons who, for personal moral reasons, studied tactical nuclear weapons in Vietnam had found the weapons were actually useful, they might not have written their report; but they would not have lied about it.

Good scientists make good advisers. Their methods of thinking about science are the most verifiable, falsifiable, and mutually understandable that humanity has ever come up with. They are drawn toward certainty but are wary of it because they know at any minute they might be wrong. They understand the concept of basing arguments, not on belief, but on evidence. And since the whole enterprise of finding the truth depends on telling it, then about their clear and beautiful science, as Prof. Y says, scientists tell the truth. When the country faces decisions about necessarily imprecise, shades-of-gray policies, it should have some truths at hand.

EPILOGUE

Outcomes and Updates

Deaths

Luis Alvarez (age seventy-seven) died in 1988. Since 1999 William Nierenberg (age eighty-one), Marshall Rosenbluth (age seventy-six), Henry Kendall (age seventy-two), Sam Treiman (age seventy-four), and Gordon MacDonald (age seventy-two) have all died.

The Manhattan Project Physicists

Enrico Fermi, Robert Oppenheimer, and Isidor Rabi—minor gods to their students—were not Jasons themselves but were Jasons' teachers. They, along with Edward Teller and Hans Bethe, had distinguished careers both in physics and in advising successive administrations on nuclear policy. Fermi died of cancer a few months after the Oppenheimer hearings, at age fifty-three. Oppenheimer died of cancer in 1967, at age sixty-three. Rabi died in 1988 at age ninety. Teller died in 2003, at age ninety-five. In 2000, at age ninety-four, Bethe wrote a letter urging the president to stop missile defense tests; in 2001 he wrote a letter with the same intent to the Congress. He died in 2005, at age ninety-eight.

The present board of sponsors of the Federation of American Scientists includes Val Fitch, Murph Goldberger, Charles Townes, Herb York, and three other Jasons or ex-Jasons. Its board of directors includes Richard Garwin and Steven Weinberg. FAS calls itself "the oldest organization dedicated to ending the worldwide arms race and avoiding the use of nuclear weapons for any purpose."

John Wheeler

In 1973 John Archibald Wheeler cowrote the standard textbook on general relativity. In 1976 he left Princeton, where he'd been since 1938, to go to the University of Texas at Austin; ten years later he came back to Princeton. In 1998, at age eighty-seven, Wheeler wrote—with the help of his former student and colleague Kenneth Ford—a readable autobiography called *Geons, Black Holes, and Quantum Foam.*

In 2001, when I interviewed him, he was about to turn ninety. I had an odd problem with him. I'd ask a question, and he'd stare up at the ceiling, motionless and silent as though someone stopped the movie, for so long I'd start thinking about resuscitation. Then finally, with little wheezy hums, he'd ease into speech. His answers were nearly always tangential to the questions, stories that would lead to other stories and then just end. *After all, he's ninety,* I thought. Not until later, when I was transcribing the interview, did I realize that the stories, put together, made up a logical, relevant, multifaceted answer to the question. I described Wheeler's conversational method to the Jasons who knew him. "He was always like that," they'd say. "He's no different."

Wheeler and his wife spend summers in the sharp light and air of High Island, Maine; winters are in Princeton. One Jason told me a story about Wheeler's having designed the Institute for Defense Analyses' logo, the stylized Greek temple that inspired Mildred Goldberger to rename Project Sunrise, Jason. Wheeler said this story rings no bells; his coauthor, Kenneth Ford, doubted the story's truth: "If one thing characterizes John Wheeler, it's a flair for the dramatic, reaching for something out of the ordinary. I just feel that if he had designed the IDA logo, it would have been more memorable."

Charles Townes

Charles Hard Townes came from South Carolina and did graduate work at Caltech; after working at Bell Labs on radar and at Columbia University on microwaves and molecules, he put in his two years at IDA, during which he helped set up Jason, and then for a few years he was provost and physics professor at MIT. In 1967 he went to the University of California, where he was made a University Professor—a title given rarely and only to eminent scholars—meaning that Townes belonged not to any one department or school but to the whole multicampused university. He moved specifically to the Berkeley campus and has been there ever since. "I guess I'm in a rut," he said. "It's a pretty good rut. As a place to live, it's great, and it's an excellent university." In 2002, at eighty-seven years old, he went to his office every day.

His Nobel Prize was for his part in inventing that ultimate technology, the laser, but he said that he'd rather do pure scientific research. He talks at the northern rate, but his sentence rhythms and choice of detail are pure South Carolina. "Now, science and technology are intimately very closely connected, and I wouldn't want to separate them. There are people who take great pleasure in making something that's going to be used immediately. That's important, and I must say I have great pleasure in seeing how useful the laser has become, particularily if some friend tells me it saved his eyes—well, that's very emotional. But I much prefer to just try to understand things. And as I say, explore the universe: the minuscule universe, tiny little atoms and parts of atoms and parts of parts of atoms; as well as the wonderful universe as a whole, an enormous thing that's just bigger than we can imagine, stars and supernovae and galaxies and supergalaxies, you might say, and the beginning of time."

Townes began his career with the same preferences: he remembers coming home from college, taking a physics text into the woods, and sitting on a rock overlooking a stream—he remembers "just exactly how it looked," he says. "I read for the first time about special relativity, and that was just a tremendous experience, to see what could be found out by reason, and a whole new view of what space and time were like."

Herbert York

In 2001 Herb York underwent chemotherapy for acute leukemia. "I watched 9/11 from my hospital bed," he says. Afterward he went dead bald, but now he has hair again. He's written several books; one is about Oppenheimer and Teller and the decision to build the hydrogen bomb; another is an autobiography called *Making Weapons, Talking Peace*. He is retired and lives on a cliff in La Jolla, in a house called Casa de los Amigos: the front gate has a little hole for the dog to stick his nose out of; through the gate is a little courtyard full of flowers and cacti, and York yells from inside the house, "You're in the right place!" The side of the house opposite the courtyard is all windows and looks out at that ocean that seems to go right on up into the sky.

York is the prototype of the Manhattan Project scientists' pragmatic moral balances. He supervised the improvement of hydrogen bombs; he worked with Edward Teller and liked him. He also repeated what he said several times in his books, that the spectacle of a nuclear bomb explosion is nothing compared to the knowledge of its effects. In 2000 President Clinton gave York, along with Sid Drell, the Enrico Fermi Award for their help in "arms control policy under four Presidents." The same year York and Murph Goldberger, both seventy-nine years old, wrote an editorial for *The Los Angeles Times* called "Star Wars II: Can We Be Protected by a Pig in a Poke?"

Project Bassoon/Shelf/Sanguine/Seafarer/ELF

In 1959 Nicholas Christofilos came up with the grand idea of communicating with submarines with extra-low frequencies and extra-long wavelengths, an idea that was eventually called Project Bassoon. Over the next decades the project "got smaller," says Jack Ruina, who was director of ARPA in the early 1960s. "Much, much smaller. Oh, yes. Got it down to sort of a realistic scale. Which was still gigantic."

So Bassoon in 1960 evolved into Shelf, and Shelf in 1972 into Sanguine. Sanguine was to transmit at a frequency of 45 hertz or a wavelength of 4,140 miles. The antenna was to be several hundred miles of cable, buried underground in Clam Lake, Wisconsin, whose underlying rock had the proper conducting properties. Sanguine evolved into Seafarer, and Seafarer into ELF. By 1989 ELF was apparently working—"fully operational" is the navy's term—at a frequency of around 76 hertz, a wavelength of twenty-five hundred miles, and an antenna that was eighty-four miles of wires strung on poles around Clam Lake and Republic, Michigan. It could send a three-letter message in four minutes.

During the 1970s the Defense Department and various consultants studied the effects on living things of electromagnetic fields this strong. The studies are exhaustive—though according to the National Academy of Sciences, also incompetent—and found no measurable harm being done to dragonfly larvae, fish, soil amoebas, slime molds, earthworms, and monkeys, among others. The citizens of Wisconsin didn't much believe these studies and held demonstrations and wrote letters for years. Murph was dismissive: "There was a hell of a lot of flap about the people in Wisconsin didn't want it, they were afraid it was going to kill their cows. Which is about as probable as getting cancer from high-tension wires."

As of 2001, a U.S. senator from Wisconsin said ELF was costing $14 million every year. In 2003 the senator reintroduced a bill—which he said he'd been introducing for much of the previous decade—to have ELF killed. In September 2004 the navy shut ELF down. The navy's press release said that ELF was no longer necessary because of "improvements in communication technology," meaning they could now do the same thing less heroically with shorter wavelengths; and because of "the changing requirements of today's navy," meaning Russian submarines aren't the worry they used to be. "This is a Cold War system," a former assistant secretary of defense told National Public Radio, "and the Cold War is over."

Murph Goldberger

I wrote Marvin L. Goldberger an e-mail asking if I could talk to him. He wrote back: "If you call home, ask for Murph Goldberger, or my wife might think you're a hustler and hang up on you. Murph G." Murph is known so universally as Murph that he's like a rock star or supermodel and almost needs no last name. He went to the same high school in Youngstown, Ohio, that John Wheeler did, and he had the same math teacher; Wheeler said her name was Lida F. Baldwin, Murph said it was Miss Dorshuk. Much later when they were both at Princeton, they argued about Vietnam so fiercely they both had sore feelings.

Murph stayed at Princeton for twenty-three years, then left the university and physics research to become president of Caltech. "To be honest with you, I became concerned that my rate of doing research had slipped while I was chairman," he says. "And the idea of walking around the corridors of Princeton and fading slowly into the sunset was not very attractive. I was fifty-five. It's an ideal age for a university president." After that he became director of the Institute for Advanced Studies, then dean of the Division of Natural Sciences at the University of California at San Diego. In his jobs as president, director, and dean, not to mention first chair of Jason, Murph's philosophy was "hire the best people, give them lots of money, and stand back."

He and Mildred now live, retired, in La Jolla in a house with an open, meandering layout that's perfect for parties. Murph at eighty has white hair, more of it than his contemporaries; all of his marbles, though he denies this; and a lively expression that isn't quite as friendly as I think it is. I asked him for the second time to explain his field of physics, and he said, "You asked me that the other day and you couldn't understand what I told you then. That would still be true." He doesn't mince words, and his language requires parental discretion. He has a kind of springy presence: he looks like he's about to jump up in the air even when he's having trouble getting to his feet. Murph supported the nuclear freeze movement; he was head and on the board of the Federation of American Scientists; his name appears regularly on letters cosigned by scientific eminences urging the government to sign the test ban treaty, stop investing in ABM systems, and stop developing small nuclear warheads for the battlefield.

Jasons and La Jolla

In the late 1950s the University of California opened a new branch that was called UC–San Diego but was actually situated just north of San Diego in La Jolla. Herb York, who had finished his term as director of defense research and engineering, became UCSD's first chancellor. Keith Brueckner, who had been at the University of

Pennsylvania, became the first chairman of its physics department and dean of its science and engineering school and was responsible for recruiting new faculty. At the Scripps Institution of Oceanography nearby, Walter Munk was a longtime and honored member; and William Nierenberg and Ed Frieman were successive directors. Between Scripps and Brueckner's recruiting at UCSD, a fair number of Jasons moved early or late to La Jolla and stayed there: Henry Abarbanel, Murph Goldberger, Ken Watson, Marshall Rosenbluth, Ken Case, and five or six others not here mentioned. It's one reason that Jasons cluster, not only at places like Princeton, Harvard, Stanford, and Berkeley, but also at the relatively less spectacular UCSD.

Nobel Prizes

Eugene Wigner, Charles Townes, Hans Bethe, Luis Alvarez, Murray Gell-Mann, Steven Weinberg, Val Fitch, Leon Lederman, and Henry Kendall won Nobel Prizes in, respectively, 1963, 1964, 1967, 1968, 1969, 1979, 1980, 1988, and 1990. Four other Jasons (or Jasons briefly) won Nobel Prizes, too, but I don't mention these scientists in the book and so won't here, either.

Val Fitch resigned from Jason in 1965. "I came to realize that when I was participating in Jason, I was subtracting a lot of my own time away from doing experiments," he says. "And Jason became less interesting for a number of reasons. The same jokes seemed to get perpetuated. The same people, with all the same prejudices. And it became a little ingrown. Though I think the idea is still a very good one. To me anyway, it started to get a little repetitive." Is the repetitiveness anything he'd like to explain? Fitch, an ex–Nebraska rancher who says as much as he wants to and not a word more, said, "No."

Steven Weinberg resigned from Jason in the early 1970s after the Vietnam studies because he didn't know whether he was doing any good and because he liked writing books, one of which, *The First Three Minutes,* sold like hotcakes. In the late 1980s Weinberg rejoined Jason as a senior adviser: he had joined in the first place, he says, for "a pleasurable sense of getting out of the ivory tower—that is, finding out what people were doing out there in the real world of applied science. I like to get that feeling again from time to time. Also I just like a lot of the people, and it gives me a chance to see them."

Vietnam

"Tactical Nuclear Weapons in Southeast Asia": The Jason report "Tactical Nuclear Weapons in Southeast Asia" stayed invisible from the time of its writing until 2003, when a public policy think-tank called the Nautilus Institute put up on its website the whole report, including interviews with its authors and commentary by various experts. The Nautilus Institute got a copy of the report via the Freedom of Information Act after nineteen years—that's nineteen years—of requests. Once the report became public, it made a modest impression on the news media, which saw ties to the current administration's recurring efforts to fund new kinds of tactical nuclear weapons. The senator from California read a statement on the Senate floor describing the Jason report, detailing the recent news, and ending with, "The conclusions of the Jason report are as valid, realistic and frightening today as they were in 1967."

The Electronic Barrier: The electronic barrier turned into the electronic battlefield, the modern method for carrying out nonnuclear warfare, in particular on the urban battlefield. The barrier was broken up into its concepts: the sensors; the orbiting aircraft (or "relay," says Richard Garwin); the computer to locate the sensors ("ground processing," says Garwin); and the bombers ("response," says Garwin). The concepts were then inventively and confusingly recombined. Sensors—"because of the wonderful reduction of the size and power of electronics," Garwin said—can now be their own ground processors, do their own analyses of sound, and distinguish between tanks and trucks. Sensors that could emit only beeps can now "store a half-an-hour speech and send it out when the time is appropriate," says Garwin, "like coming home at the end of the day and telling your family what happened to you." The relay to which the sensor talks is now a UAV, an unmanned aerial vehicle like the Predator or the Global Hawk, used in both Gulf wars and in Afghanistan. Because the sensors can store their stories, the UAV/relays don't have to stay in constant touch but can fly over the sensors occasionally and poll them. The UAV/relays can also carry their own sensors, radar or visual or infrared; some of their sensors can make images of moving things. The central ground processor to which the UAV/relay talks, Garwin says, "no longer has to be in the theater, it could be perfectly well in the United States or in Europe." The responses are now bombs that are guided by lasers or that seek their own targets or that have their own GPS navigation. "There's BAT," says Garwin, which has acoustic sensors "that look for the sounds from tanks." Other bombs home in on infrared light, "so they attack hot things, like decoy fires." The military calls the whole idea "sensor to shooter." A minor industry is devoted to shortening the time from one to the other.

The Electronic Barrier on the Mexican Border: A less lethal, less high-tech application of the barrier's technology was strung along the border between the United States and Mexico, to find illegal immigrants and drug smugglers. When the sensors go off, a radio dispatcher sends out immigration officers, not bombers. Jason eventually worked on this system, too; it's the one about which Freeman Dyson told the Johns Hopkins physicists and that triggered my preoccupation with the Jasons. "We were working for Customs and Immigration, I guess," Dyson said. "We were

trying to stop the drug traffic. Simply, we wanted to educate ourselves about what was going on. We talked to the border patrol,
and they said, 'Well, if you really want to see what's going on, why
don't you come out at night and see for yourselves?' It was quite
dramatic. It gives one an idea of how difficult it is to control a
border." That trip to the border was in July 1989. Along with
Dyson were other Jasons, including William Press. A mile north
of the border, in more or less open desert, the trails and roads
were strewn with several hundred sensors, "both seismic and
magnetic," Press said, "surplus from the Vietnam era. When they
were triggered by a passerby, they radioed back to sector headquarters." A sensor went off "every five or ten minutes," said Press.
At the headquarters' communication center the sensor signals
showed up on computer screens. This barrier wasn't stopping infiltration, either, but only because there were too many infiltrators.

Charles Schwartz: Charles Schwartz was awarded tenure at Berkeley and around 1970 began requiring his students to sign a Hippocratic oath pledging not to use the physics he taught them to harm
anyone. The physics department threatened to remove Schwartz's
tenure, and Schwartz abandoned the requirement. Later he
stopped teaching physics because, he said, he was only supplying
bodies to the defense contractors. Instead he taught courses on the
relationship between science, government, and society. Eventually
he stopped receiving raises because he was doing so little physics;
tenure made his position relatively secure, however, and he said
that more faculty members should use their tenure to be activists.
"I feel very good about this kind of a life," he said. Schwartz said he
survived as a radical because he's a physicist: "You just keep after it
and keep after it and you get a theory and you try it and it doesn't
work and you get another one. . . . Just this terrible grinding dedi-
cation, the keeping after it. Most physicists pursue physics like

that." In 1987 Schwartz told National Public Radio that although the Jasons claim to tell generals when weapons won't work, in effect they "just keep the Pentagon more efficient." In 1993 Schwartz became professor emeritus at Berkeley. In 2002 he put his booklet, *Science Against the People—The Story of Jason,* on the Internet.

Adaptive Optics

After adaptive optics and laser guide stars were turned over to astronomers, Claire Max and her colleagues built a working system and installed it all on a normal-sized, three-meter telescope at the Lick Observatory. "It's working," Max says, "and doing science beautifully." Since then Max has more or less sent her career off in that direction. She's moved and now spends most of her time at the University of California at Santa Cruz (UCSC), where Lick is and where she helped set up a center—of which she is now director—for further research in adaptive optics.

The center's projects are two new kinds of adaptive optics. One, called extreme adaptive optics, deforms the mirror not at the current 250 different places but at five thousand different places, correcting the tiniest distortions and allowing the kind of sharpness that could pick out tiny dim planets swamped in the light of their big bright stars. The center's other project is called multiconjugate adaptive optics. Instead of using one laser guide star to measure the turbulence in one patch of the atmosphere, it uses several, each aimed in slightly different directions. The upshot is, said Max, "you're actually taking slices through the atmospheric turbulence in all these different directions." They're effectively doing tomography, the way ocean acoustic tomography images the ocean's temperatures, but in this case, Max says, "you can reconstruct what the whole three-dimensional atmospheric turbulence looks like," allowing astronomers a larger field of undistorted view.

Since the early 1980s, when Jasons thought up the sodium laser guide star, astronomy has been on a hot roll made possible by a number of new technologies. Astronomers who had been limited by small mirrors to seeing bright things, learned to make mirrors two or three times larger and see faint things. But no matter how large the mirrors or how faint the images, astronomers had been restricted to the level of detail allowed by the atmosphere. Astronomers measure detail, or resolution, in arc-seconds; the moon is 1800 arc-seconds across; a shirt button a thousand meters away is one arc-second across. The best resolution that telescopes could get in the early 1980s was one arc-second, or one-eighteen-hundredth of the width of the full moon; astronomers used to say, "It's a one-arc-second atmosphere." The Hubble Space Telescope got so famous not because its mirror was so large—it was in fact relatively small—but at least in part because it was above the atmosphere and had a resolution of one-twentieth of an arc-second.

But space telescopes are few and ground telescopes are many. By 2005 twenty-two telescopes, sited all over the world, with mirrors of all sizes, had adaptive optics systems either in development or operating. Nine of them had sodium laser guide stars either in development or operating. Max can name every telescope and every system. The best of them has a resolution of around a hundredth of an arc-second. Astronomy websites show adaptive optics before-and-after pictures: blotches of light are sharpened into storms on Neptune; a few fuzzy blobs become hundreds of stars around a black hole; an out-of-focus asteroid turns out to be an asteroid and its tiny moon. In 2005 a European telescope fitted with an adaptive optics system confirmed the first pictures taken of a planet outside our solar system.

The astronomers building those ground telescopes are planning bigger and bigger mirrors to see farther and fainter yet. Wayne van Citters, the National Science Foundation's director of astronomy, says that most of the astronomical community believes those big telescopes are worth building only if they can incorporate adaptive optics systems. The military is using adaptive optics and laser guide stars, too, but aren't posting before-and-after pictures on the Internet; for billions of dollars, they're putting the systems on Boeing 747s that from high orbits can aim killer lasers at battlefield missiles hundreds of miles away.

Meanwhile, in yet another example of the futility of answering, once and for all, the moral questions about science's applications: the UCSC center is using adaptive optics technology in devices that overcome natural distortions in the eye to make images of a living human retina. The resolution in these images is high enough to make out individual cells and the smallest veins, and therefore to help diagnose and treat retinal disease.

Another Jason Invention

Any country that's hosted a modern war has residual, unexploded land mines. Military deminers, clearing ex-battlefields of old mines, use heavy, expensive equipment. A civilian deminer—say, a farmer clearing his field—lies down and pokes a stick into the ground until he hears a particular thunk, then prods around to see if the thing he hit is large enough to be a land mine. In 1996 DARPA asked Jason to review the new lightweight, cheap technologies that civilians could use. Over half the Jasons worked on that study. In their report, "New Technological Approaches to Humanitarian Demining," they suggested some technologies of their own.

Since most mines give off vapors of TNT, dogs can be trained to find mines reliably. Could you also train honeybees, which can smell chemicals at one part in 10 billion, to be attracted to TNT? Research in bees' sense of smell goes back at least to the 1950s, says Paul Horowitz, who led the Jason study, citing the paper "Über die

Riechschärfe der Honigbiene," which he translated as "Over the Smell-sharpness of Honeybees." Bees are trainable, but training isn't inherited, so "you have to train each bee," he said. "That's the problem with bees." Then could you genetically engineer fruit flies so they're attracted to TNT and hover over a mine as if it were an old banana? Or to be really tricky, says Horowitz, how about engineering fruit flies so they're not only attracted to TNT but also light up like fireflies and map out land mines with little sparkling clouds? "It sounds completely off the wall, but the ideas came from some card-carrying biologists," Horowitz says, "and I've run these past some additional card-carrying biologists, and they outline the steps you would do to carry it out."

More prosaically, could you build a probe that would bounce X-rays or microwaves or sound off a suspected land mine and attach to the probe a sensor that could tell by the bounce-back whether something was plastic or wood or stone or metal? Horowitz was so intrigued by the idea of an acoustic land mine detector that he got a little money from a private foundation for a student to build one and test it to see if it could find inert mines. It worked nicely in a homemade minefield outside the Harvard physics building, but in one of two follow-up tests with eight inert mines, it missed two. "It wasn't a machine we would have followed into a minefield ourselves," Horowitz says.

More Stockpile Stewardship Studies

For years, Jason has reviewed the weapons labs' programs to ensure that the country's nuclear stockpile remains capable. The most controversial of these programs—what Will Happer's called "Christmas presents" to the labs for supporting the Comprehensive Test Ban Treaty—has been the attempt to create fusion with lasers at the National Ignition Facility, or NIF. In 2000 the U.S. General Accounting

Office's report on NIF, "Management and Oversight Failures Caused Major Cost Overruns and Schedule Delays," said that NIF, with a new completion date of 2008, was running six years behind schedule and at $3.9 billion, was $1.8 billion over budget.

By the end of 2004, Congress was "extremely concerned" by a proposal that NIF's completion date be moved from 2010 to 2014, and asked that Jason be commissioned to review the outlook for NIF's success. Jason began the third of its studies on NIF early in 2005, breaking its yearly routine by working through the spring. At the end of June 2005, Jason's report became public. NIF will probably make its first attempt at fusion by the 2010 deadline, the report said, but the success of that attempt, "while possible, is unlikely."

Jason recommended a series of managerial and scientific steps that would keep NIF on track. Whether Congress continues funding NIF is, as of this writing, unknown; in early July, the Senate voted to stop funding its construction. But back in 2003, when Congress asked for a review of another stockpile stewardship program in trouble, Jason said that the program was a pillar of stockpile stewardship and recommended a road map. In 2005 Congress added $75,000,000 to the program's budget.

Another Detected Lemon

In 2003 DARPA was deciding whether to fund work on a so-called hafnium bomb. A hafnium bomb was a new kind of nuclear weapon, made by hitting the radioactive form of the element hafnium with X-rays. Hafnium would soak up the X-rays' energy, then release it all at once in an explosion. A hafnium bomb was either cutting-edge physics or snake oil. Jason studied an experiment verifying its possibility, found it implausible in theory and unconvincing in practice, and suggested a more sophisticated repeat of the experiment. Several sophisticated experiments later and still no bomb, DARPA budgeted $4 million for the hafnium bomb anyway. In 2004 Congress cut the budget by $4 million, citing "expert technical opinions." "What we're best at," says Freeman Dyson, "is shooting down stupid ideas."

Freeman Dyson

Freeman Dyson at age eighty has a Duke of Windsor haircut, gray with a little dark still in it. His ears stick out; he has a long face and round, cloud-blue eyes; and sometimes he looks at you so directly, it's disconcerting. His expression is anticipatory—it's either anxiety that you'll say something egregiously stupid, or hope that you'll say something delightfully original, one or the other. When you walk into his office for an appointment, he acts amazed. When he smiles, he looks stunned with joy.

Dyson is British. His sister, Alice, remembers him as a small child sitting surrounded by encyclopedias and sheets of paper on which he was calculating things. At public school he did math. In 1943 he worked as a civilian doing operations research for the Royal Air Force's Bomber Command; he was nineteen years old. "Everybody was young," he says. "That was a children's war."

In 1947 he got a fellowship for graduate school at Cornell, where he became Hans Bethe's student. Bethe eventually recommended that Dyson spend time with Robert Oppenheimer at the Institute for Advanced Study. Bethe wrote to Oppenheimer, "I can say without reservation that he is the best I have ever had or observed." In 1948 Dyson drove from Cleveland to Albuquerque in a secondhand Oldsmobile with Richard Feynman, also at Cornell and fresh out of the Manhattan Project. "As we drove through Cleveland and St. Louis, [Feynman] was measuring in his mind's eye distances from ground zero, ranges of lethal radiation and blast and fire damage," Dyson wrote. "I felt as if I were taking a ride with Lot through Sodom and Gomorrah."

Dyson never finished his Ph.D. He says that doctoral students end up middle-aged, overspecialized, unprepared for the outside world, trapped, depressed, discouraged, and mentally deranged; not having a Ph.D. is "a badge of honor." He is nevertheless famous for some of the most fundamental work on quantum field theory—some physicists say he should have won the Nobel Prize—and was offered several jobs on the strength of it. In 1953 he went to the Institute for Advanced Study and has stayed there to this day. He has helped design nuclear reactors that produce the radioactive forms of atoms used in medical tests. He designed and forcefully backed a nuclear-powered space ship that, because it would have sprayed radioactive fallout behind it, luckily never got off the ground; his son, George Dyson, wrote a history of it called *Project Orion*. Dyson wants badly to go to space and has come up with several schemes—one a Jason project—for doing so. He may have joined Jason at the beginning—quasi-official membership lists record his name as of 1959 or 1960—but he says he spent no summers there until 1965.

"I have two skills," he says. "One is calculating and the other is writing English prose." His calculations strike awe into Jasons: the typical story is that someone poses a mathematical problem, Dyson answers it immediately, and the rest of the Jasons go home and think about it for hours and still don't understand his answer. His English prose is impeccable and flat-out charming. He's written many books of reviews and essays whose subjects are surprisingly varied and whose content is rich and provocative, to which he introduces the reader through stories from his own life. He's thought hard and creatively about science and morality, science and war, science and the government. Nevertheless, he's never been a white-collar Jason, has not even served on the program committee: "I never went around in Washington talking to the sponsors.

I've obviously no talent for it, I suppose. The fact is, when I'm on a committee, I usually go to sleep. Some people are just no good at committees."

Dyson talks with long pauses, not because he has trouble with words or names—he doesn't—but because he seems to write the sentences first in his head, getting the ideas into the right words, the words into the right order. His sentence inflections fall at the end and sound a little sad. He says he's "fading out" of Jason. "I don't know how long I'll hang on there," he says, though to my certain knowledge he's been saying that for the last ten years. "So I probably will stay until I'm kicked out." He likes Jason better than the Institute for Advanced Study, he says, because Jason is like being back in graduate school at Cornell, with Bethe and Feynman and the other Manhattan Project veterans, working together on the same problems. "It's a much more sociable life at Jason than I have here," he says. "I'm not doing anything here. All I do here is write book reviews and prefaces. I write prefaces for other people's books." He gave one of his whispered laughs, apparently at the outcome of human endeavor. "Here, I'm in isolated splendor."

The Retirements of Garwin and Drell

Sid Drell was born in 1926. He has now retired from Stanford but remains on campus as a senior fellow at the Hoover Institution, a politically right think-tank of which, he says, he is not a typical member. Richard Garwin was born in 1928. He retired from IBM, became a senior fellow of the Council on Foreign Relations, and retired from there, too. Neither one, as far as I can tell, has changed his habits or activity levels at all, and both will apparently go on fighting nuclear proliferation, ignorance, and illogic through all eternity.

Notes

A rule of nonfiction writing, which I have followed, is that any given piece of information needs more than one source. For the sake of brevity, however, and with some exceptions, I have listed only those sources that are the most comprehensive, authoritative, and/or reliable.

Most of the book's information and the majority of quotes come from my interviews and are not noted in this section; a list of those interviews and their dates appears in the Sources.

Quotes from correspondence with me and all other sources are noted below. Quotes from unnamed parties who did not agree to be named are left anonymous and kept to a minimum.

AIP, unless otherwise specified, refers to the Oral History Project at the American Institute of Physics, Center for History of Physics, at the Niels Bohr Library, in College Park, Maryland.

INTRODUCTION: Jasons

xiv *I did find a remarkable archive of transcribed interviews*: AIP.

xxiv *Jasons are now paid . . . budget is $3.5 million*: Robert Henderson, e-mail to author, October 4 and December 9, 2004.

xxvii *For many of these anonymous Jasons . . . ensuing personal attacks*: Will Happer, e-mail to author, May 1, 2000.

xxviii *One told a story about a Nobel Prize–winning Jason . . . yelling . . . "Helllppp!"*: Edward Frieman, letter to author, October 28, 2004.

xxx *"We are scientists second . . . knowledge implies responsibility"*: Dyson, *Disturbing the Universe*, 6.

CHAPTER ONE: **The Bombs**

1 *The idea that curiosity leads to disaster . . . medieval Christian theologians, considered* curiositas *a vice*: Shattuck, *Forbidden Knowledge,* 13–47.

1 *By the time the war ended, . . . chemical warfare had killed or injured a million soldiers*: Rhodes, *Atomic Bomb,* 91–95; "A Short History of Chemical Warfare During World War I," online at http://www.mitretek.org/ home.nsf/homelandsecurity/WWIChemHistory.

3 *"Even before the end of our first year at Los Alamos, . . . oh God are we doing right?"*: Elsie McMillan, "Outside the Inner Fence," in Badash, Hirschfelder, and Broida, *Reminiscences,* 43.

3 *In 1943 Herbert York was a graduate student in physics . . . cursing one's mother"*: York, *Making Weapons,* 6, 14–15.

3 *William Nierenberg was a graduate student . . . a different method*: William Nierenberg, interview by Finn Aaserud, AIP, February 6, 1986, 27.

3 *Mildred Ginsberg . . . also at Chicago*: Mildred Goldberger, e-mail to author, July 16, 2003.

4 *John Wheeler, a physicist at Princeton, . . . worked on manufacturing plutonium*: Wheeler, *Geons,* 36–37.

4 *Edward Teller, . . . the atmosphere and planet were safe*: Rhodes, *Atomic Bomb,* 418–19; Serber, *Los Alamos Primer,* xxxi.

4 *The SEDs, though in the army . . . "Even the sergeant broke down and dismissed them"*: Hoddeson, *Critical Assembly,* 97–98; Bernice Brode, "Tales of Los Alamos," in Badash, Hirschfelder, and Broida, *Reminiscences,* 145.

5 *A civilian scientist intervened . . . latrine cleaning*: Kistiakowsky, "Reminiscences of Wartime Los Alamos," in Badash, Hirschfelder, and Broida, *Reminiscences,* 57.

5 *Saturday morning inspections "became devoid of spit and polish"*: Val L. Fitch, "The View from the Bottom," in Wilson, *All in Our Time,* 44.

5 *Fitch helped measure . . . stood outside, and watched the test*: Val Fitch, e-mail to author, October 19, 2004.

5 *The uranium bomb was too simple to need testing*: Badash, Hirschfelder, and Broida, *Reminiscences,* xviii.

5 *The flash of light from the explosion . . . "famous mushroom cloud rise in the morning sky"*: Fitch in Wilson, *All in Our Time,* 45–46.

5 *"Now I am become Death, the destroyer of worlds"*: Quoted in Rhodes, *Atomic Bomb,* 676.

6 *"Suddenly, . . . A new thing had just been born"*: Ibid., 672.

6 *"a good shot from the bottle"*: Fitch in Wilson, *All in Our Time*, 46.

6 *Enrico Fermi . . . equivalent to ten kilotons*: "Trinity Test, July 16, 1945, Eyewitness Accounts—Enrico Fermi," U.S. National Archives, Record Group 227, OSRD-S1 Committee, Box 82, Folder 6, "Trinity." Online at http://www.dannen.com/decision/fermi.html.

6 *Luis Alvarez . . . saw the Trinity test from a B-29 at twenty-four thousand feet*: "Trinity Test, July 16, 1945, Eyewitness Accounts—by Luis Alvarez," online at http://www.dannen.com/decision/alvarez.html.

6 *Another physicist at the Radiation lab . . . "was to make sketches of the mushroom cloud"*: Wolfgang Panofsky, e-mail to author, May 4, 2004.

7 *Oppenheimer said that the knowledge . . . "a knowledge which they cannot lose"*: Oppenheimer, "Physics in the Contemporary World," 202.

8 *Less than ten years before, . . . Frisch thought a good name for the process would be* fission: Frisch, *What Little I Remember*, 113–17.

8 *"The bomb was latent in nature as a genome is latent in flesh"*: Rhodes, *Atomic Bomb*, 379.

9 *Oppenheimer told the Los Alamos . . . "lights and its values"*: Smith and Weiner, "Robert Oppenheimer," 15.

9 *The historian Daniel Kevles wrote . . . "control of atomic energy"*: Kevles, *Physicists*, 337.

10 *In June 1945, a month before Trinity, . . . "the existence of nuclear weapons"*: Smith, *Peril and Hope*, 41–53, 71, 371–83; Kevles, *Physicists*, 334–36.

10 *Another attempt came the day after Trinity . . . "international controls over atomic power developments"*: Smith, *Peril and Hope*, 54–56, 93.

10 *Yet another attempt came the next month, . . . giving nuclear researchers a venue for international conversation*: Ibid., 128–73, 271–75; Kevles, *Physicists*, 349–52.

11 *And again: four months after Hiroshima . . . testifies on just about anything in national policy that's related to science*: Smith, *Peril and Hope*, 203, 279–80. Current information about the Federation of American Scientists comes from its website, at http://www.fas.org.

11 *The physicists could not control . . . "physicists were popular at dinner parties"*: Kevles, *Physicists*, 367–69. The Oppenheimer quote is from Oppenheimer, "Physics in the Contemporary World," 200. The Wheeler quote is from interview by Finn Aaserud, AIP, May 23, 1988, 41. The Townes quote is from Townes, *How the Laser*, 49.

12 *"We all felt that, like the soldiers,"... overflowing with students.*
Bethe, "Observations." Written in 1954, this work was declassified in 1980
and is available on the website of the Federation of American Scientists, at
http://www.fas.org/nuke/guide/usa/nuclear/bethe-54.htm.

12 *Val Fitch still had to finish college... "So we were Fermi's first two
theoretical students."* York's personal information is from York, *Making
Weapons*, 31. Nierenberg's biography is on the website of the Scripps Institution of Oceanography, at http://scilib.ucsd.edu/sio/archives/siohstry/
wan-biog.html.

13 *One of Hans Bethe's new graduate students... "comradeship and
deep happiness"*: Dyson, *Disturbing the Universe*, 20–21, 51–53.

14 *Several years earlier, back at Los Alamos, Fermi had proposed... too
busy with the fission bomb to worry about a fusion bomb*: Hoddeson, *Critical
Assembly*, 44–45; Rhodes, *Atomic Bomb*, 374, 754.

14 *In 1946 Winston Churchill... "control from Moscow"*: Churchill's
speech is online at http://www.historyguide.org/europe/churchill.html.

15 *Teller,... Luis Alvarez agreed with Teller*: Wheeler, *Geons*, 280;
Rhodes, *Atomic Bomb*, 764–68; and Bethe, *Road from Los Alamos*, x.

15 *Other Manhattan Project scientists... "It is necessarily an evil
thing considered in any light"*: Rhodes, *Atomic Bomb*, 769; document online at http://www.pbs.org/wgbh/amex/bomb/filmmore/reference/primary/
extractsofgeneral.html.

15 *John Wheeler, now back at his faculty job... being run with graduate
students and new Ph.D.'s*: Wheeler, *Geons*, 218–19, 227.

16 *Both Richard Garwin... the design "that Teller wanted," he said, but
that worked as Garwin had set it up to work*: Richard Garwin, e-mail to author, May 23, 2003; Garwin, interview by Finn Aaserud, AIP, February 4,
1987, 12–13; Garwin, interview by Finn Aaserud, AIP, October 23, 1986, 18.

17 *Ed Frieman saw three of the shots... "standing on end with tons of
roiling mud"*: Ed Frieman, interview by Naomi Oreskes and Ron Rainger,
February 17, 2000, transcript, 2, online at http://www.heinzctr.org/Programs/
SOCW/interviewers_colloquia.htm. This interview is no longer accessible
on this website or available on any other.

18 *The Franck Report had a preamble... "the use to which mankind
had put their disinterested discoveries"*: Smith, *Peril and Hope*, 372.

19 *In the spring of 2000 he gave a speech... "preserving your society
against ignorance and unreason"*: Richard Garwin, Berkeley Physics

Graduation Speech, May 17, 2000, online at http://www.fas.org/rlg/ 000519-berkeley.htm.

CHAPTER TWO: Jason Is Born

20 *Herb York got his doctorate in 1949, ... "or the highest yield-to-weight ratio"*: The Livermore history comes from its website, at http://www.llnl.gov/llnl/history/lawrence.html and http://www.llnl.gov/ timeline/50s.html. York's quote is from York, *Making Weapons*, 75.

21 *"Everyone who was at all knowledgeable, ... a target sixty-five hundred miles away are virtually identical"*: Herb York, e-mail to author, August 19, 2002.

21 *The weight of hydrogen bomb warheads at the time varied, but they were closer to a thousand pounds than to 31*: Herb York, e-mail to author, October 25, 2004.

21 *This disparity in size . . . the system was so highly classified, no one could say so*: Charles Maier in Kistiakowsky, *Scientist*, xxxiii; York, *Making Weapons*, 100–11.

22 *"Arms builders everywhere," York said . . . save the republic from imminent disaster"*: "Interview with Herbert York," online at http://www.gwu.edu/~nsarchiv/coldwar/interviews/episode-8/york3.html.

23 *A month and a half after Sputnik, . . . Its chair was Isador Rabi, and its members included Hans Bethe*: Garwin quote is from Richard Garwin, interview by Finn Aaserud, AIP, October 23, 1986, 36; York quote is in York, *Making Weapons*, 105. The number of scientists in the first PSAC is from York, *Making Weapons*, 106 and footnote. PSAC's first questions are from Charles Maier in Kistiakowsky, *Scientist*, lvi.

23 *The secretary had in mind "anti-missile missiles and outerspace projects," but wouldn't rule out "highly speculative types" of weapons*: Neil McElroy to John Stennis (chairman of the Subcommittee on Military Construction, Senate Armed Services Committee), January 1958; Neil McElroy testimony, to House Committee on Appropriations, 85th Cong., 2nd sess., January 15, 1958.

23 *His decision to join the Defense Department, . . . "enhancing a professional reputation were not among them"*: Townes, *How the Laser*, 135.

24 *"The power of the purse' is just about all the power there is"*: York, *Making Weapons*, 168.

24 *And so for the next decade, . . . it's nearly vertical*: Advanced Research Projects Agency, *1958–1974*, fig. 1–2.

24 *The money going into research doubled; . . . and awarded half the doctorates*: Charles Maier in Kistiakowsky, *Scientist*, lxii–lxiii; Kevles, *Physicists*, 386, 388.

25 *Wheeler was worried. . . . "could be useful"*: Wheeler, *Geons*, 271.

25 *On the other hand, Wheeler . . . disapproval, he said, was "deeply troubling*: Ibid., 199.

26 *"Actually, we all know each other," . . . It's just us boys"*: Boehm, "Pentagon," 160.

26 *"You would find a larger percentage of people . . . than you would along the River Charles"*: John Wheeler, interview by Finn Aaserud, AIP, November 18, 1988, 16–17.

27 *The way to get new people involved . . . National Advanced Research Projects Laboratory*: Wheeler, quoted in Aaserud, "Sputnik," 197.

27 *It went to just about everybody . . . research initiation laboratory, a summer study*: Aaserud, "Sputnik," 197–212. Herb York interview by Finn Aaserud, AIP, February 7, 1986, 4.

27 *Summer studies had been invented . . . "Summer studies and some are not"*: Richard Garwin, interview by Finn Aaserud, AIP, October 23, 1986, 33–34; Townes, *How the Laser*, 139–140; Zacharias, quoted in Kevles, *Physicists*, 376.

28 *The Wheeler-Wigner-Morgenstern summer study . . . "a good idea and had money"*: Cost is in Aaserud, "Sputnik," 212; Herb York, interview by Finn Aaserud, AIP, February 7, 1986, 5.

29 *The scientists invited to Project 137 were younger than Wheeler . . . All likely had top secret clearances*: Scientists' ages are in Aaserud, "Sputnik," 215. Affiliations and fields of the scientists are in U.S. Department of Defense, *Identification of Certain Current Defense Problems*, 137–38. Clearances are in ibid., 16.

29 *The briefings that most impressed the participants . . . in missiles and in radar and electronic equipment*: Ibid., 19–21.

30 *Wheeler remembered briefings on guerrilla warfare . . . "the ability of an insect to detect a smell"*: John Wheeler, interview by Finn Aaserud, AIP, May 23, 1988, 157–58.

30 *Watson thought the briefings . . . "were not very familiar with much of this"*: Kenneth Watson, e-mail to author August 22, 2003.

30 *The resulting report, "*Identification of Certain Current Defense

Problems" . . . *"into which we are inexorably heading"*: The conversion of radios is in Department of Defense, *Identification of Certain Current Defense Problem*, 83–85; the torpedo is at 102–17; quote is at 18–19.

31 *Under all the briefings and calculations . . . much like Wheeler's original possibilities*: Wide support for a research initiation lab is reported in Wheeler, *Geons*, 282; see also Aaserud, "Sputnik," 217–18. Forms of the lab are in U.S. Department of Defense, *Identification of Certain Defense Problem*, 24–28.

31 *York asked first Wheeler . . . direct the laboratory*: York, *Making Weapons*, 152–53.

32 *If Wheeler wouldn't stay . . . Wigner and Morgenstern*: Aaserud, "Sputnik," 224.

33 *In what York called "a process of serendipity" . . . "what we now call Jason"*: Herb York, interview by Finn Aaserud, AIP, February 7, 1986, 4–5.

34 *In 1960 the average public school teacher . . . between $12,000 and $15,000*: Teachers' salaries are from the National Center for Education Statistics, Digest of Education Statistics Tables and Figures, 1999 http://nces.ed.gov/programs/digest/d99/d99t081.asp. Congressmen's salaries are from Burtless, *How Much?*, fig. 1. Physicists' salaries are from Kevles, *Physicists*, 388.

35 *Moreover, at Columbia Townes had met . . . "and thought, 'What a fine thing it is'"*: Charles Townes, interview by Finn Aaserud, AIP, May 20–21, 1987, 77.

36 *IDA, which also hosted Project 137, . . . the Joint Chiefs of Staff and the secretary of defense*: Office of Technology Assessment, *History of the Department of Defense*, 2–3, 26.

36 *Stern, though he agreed about the conversation . . . "You know what I mean?"*: Marvin Stern, interview by Finn Aaserud, AIP, May 1, 1987, 45.

37 *"When I interacted with Townes . . . potential significance of something like this"*: Ibid., 52.

37 *So Townes talked to his superiors . . . thought it was a good idea*: Townes, *How the Laser*, 139.

37 *Townes told Goldberger . . . "rather than selling something to the government"*: Charles Townes, interview by Suzanne Riess, University of California, Berkeley.

38 *the contract between ARPA and IDA was for $250,000*: MacDonald, "Early Years," 4.

38 *In addition to setting up a support channel . . . "to see that they were*

more independent": Townes, interview by Suzanne Riess, University of California, Berkeley.

39 *Besides July-August-September-October-November . . . "after their family dog"*: York, draft of *Making Weapons*, September 23, 1985, 63; possession of Marvin Goldberger.

39 *The authors of the official . . . "namely, 'the golden fleece.'"* U.S. Department of Defense, *Advanced Research Projects Agency, 1958–1974*, IV–34.

40 *"Some of them were young . . . the kind of clearance we felt important"*: Townes, interview by Suzanne Riess, University of California, Berkeley.

41 *"I would say that is undoubtedly the most important single thing that I did"*: Ibid.

CHAPTER THREE: **The Glory Years**

42 *Jason was born . . . ARPA historians called "the golden era"*: U.S. Department of Defense, *Advanced Research Projects Agency, 1958–1974*, I–4.

42 *Robert Oppenheimer, asked . . . "only after you have had your technical success"*: U.S. Atomic Energy Commission, *Matter of Oppenheimer*, 79–81.

43 *Watson, Brueckner, and Murph . . . the first steering committee*: Aaserud, "Sputnik," 229.

43 *Recruiting them turned out to be easy . . . "It was an honor to be asked"*: The number of invitees comes from Aaserud, "Sputnik," 233. Keith Brueckner, interview by Finn Aaserud, AIP, July 2, 1986, 8.

43 *"would have been hard-pressed to recruit replacements of equal calibre"*: U.S. Department of Defense, *Advanced Research Projects Agency, 1958–1974*, IV–34.

43 *Among those at this first meeting . . . and Ed Frieman*: Aaserud, "Sputnik," 203n, 233.

44 *He was a Russian . . . became Eisenhower's second science adviser*: Biographical details from Kistiakowsky, *Scientist*, v, and from "Biography of George B. Kistiakowsky," online at http://www.aip.org/history/ead/harvard_kistiakowsky/19990028.html. On the Los Alamos experience, see Kistiakowsky, "Reminiscences of Wartime Los Alamos," in Badash, Hirschfelder, and Broida, *Reminiscences*, 57.

44 *"Thursday, December 17, 1959 . . . just like other luncheons"*: Kistiakowsky, *Scientist*, 200–201.

45 *"minimum expenditures will be made for computers [and] assistants"*: A. W. Betts (director of ARPA) to Garrison Norton (president, Institute for Defense Analyses), memorandum on Project Sunrise, ARPA Project Assignment, January 1, 1960.

45 *ARPA ended its project assignment . . . "be pursued vigorously"*: Ibid.

46 *So when Townes called Drell . . . "one's honored to be called by Charlie Townes"*: Sidney Drell interview by Finn Aaserud, AIP, July 1, 1986, 11.

46 *"You can't be tight . . . Goldberger has a loose style anyhow"*: Sam Treiman interview by Finn Aaserud, AIP, December 18, 1986, 28.

46 *Townes said they were "somewhat figureheads," though reassuring and encouraging*: Charles Townes, interview by Finn Aaserud, AIP, May 20–21, 1987, 120.

47 *"It always happens with old boys' networks . . . all human traits"*: Sam Treiman interview by Finn Aaserud, AIP, December 18, 1986, 31.

47 *New members weren't official . . . "and they were pretty sure to have you"*: David Katcher, interview by Finn Aaserud, AIP, April 16, 1986, 32.

47 *"All sorts of ways . . . Jason's possible contribution would develop"*: Ibid. 44.

48 *"You were never, never directed"*: Sam Treiman, interview by Finn Aaserud, AIP, December 18, 1986, 34.

48 *"everybody got reports . . . they worked together"*: David Katcher, interview by Finn Aaserud, AIP, April 16, 1986, 35.

49 *"credited with major contributions to ballistic missile defense"*: Advanced Research Projects Agency, 1958–1974, I–10.

49 *Defender had "a slightly flaky" . . . roughly half of ARPA's budget*: Quote on Defender's flakiness is from ibid. IV–22; Defender costs are from ibid., IV–21.

49 *"resulted in major contributions"*: Ibid. IV–35.

49 *One of these contributions . . . Around twenty Jasons came*: MacDonald, "Early Years," 5.

51 *Jasons met on campus and lived in nearby rented farmhouses*: Francis Low, interview by Finn Aaserud, AIP, April 29, 1986, 6.

52 *"government simply didn't . . . to do an experiment"*: Hal Lewis, interview by Finn Aaserud, AIP, July 6, 1986, 41.

52 *"on the rolling hills of Stanford . . . high energy physicists"*: Marvin Goldberger, interview by Finn Aaserud, AIP, February 12, 1986, 11.

52 *"an old radar set and sent . . . hillside nearby"*: Hal Lewis, interview by Finn Aaserud, AIP, July 6, 1986, 41.

53 *"We worked on it over and over again"*: Keith Brueckner, interview by Finn Aaserud, AIP, July 2, 1986, 12.

53 *"The particle beam was a creation . . . at the same time"*: Quote about Christofilos's character, *Advanced Research Projects Agency, 1958–1974*, IV–23. Quote about Buck Rogers's image and the endurance of Project Seesaw, ibid., IX–31. Quote about Christofilos's successive solutions, ibid., IX–32. Quote from York, ibid., IV–23.

54 *Nick Christofilos was born . . . a job at Brookhaven that, in 1953, he took*: The story of Christofilos's three letters is in York, *Making Weapons*, 129–30; see also Melissinos, "Christofilos"; "Volatile Scientist," *New York Times*, March 19, 1959, 16; and obituary, "Nicholas C. Christofilos," *Physics Today*, January 1973, 109–12. Courant's account and quote are in Ernest D. Courant, "Biographical Memoirs: Milton Stanley Livingston May 25, 1905—August 25, 1986," online at http://bob.nap.edu/html/biomems/mlivingston.html.

55 *Christofilos designed a fusion machine, . . . prove it wouldn't work someday*: Bromberg, *Fusion*, 122, 201–204.

56 *The Argus experiment was actually three . . . "It almost seemed impudence"*: Yield of explosions is from Federation of American Scientists, "United States Nuclear Tests—By Date," 8, online at http://www.fas.org/nuke/guide/usa/nuclear/209chron.pdf. The description of test results is in Christofilos, "Argus Experiment," 1147–48. Frying satellites are in *Jane's Defense Weekly*, October 23, 2002, 20–23. The leak to *The New York Times* is in Kistiakowsky, *Scientist*, 72, and York, *Making Weapons*, 149–50. *The New York Times* articles appeared on March 19, 1959, 1; March 22, 1959, E6; March 26, 1959, 1; and April 30, 1959, 14. The quote from "Veil Around the World" is from *Time*, March 30, 1959, 70–71.

56 *"One of Jason's big jobs . . . wouldn't work"*: Matthew Sands, interview by Finn Aaserud, AIP, May 4, 1987, 72.

57 *The first name for the scheme . . . before the first Argus shots*: Merrill, "Some Early," 360. Project 137 mention of Bassoon is in U.S. Department of Defense, *Identification of Certain Current Defense Problems*, ii, 9.

57 *Two years after Project 137 . . . through the other end*: Christofilos, "ELF—Communication System," 145.

58 *The loop in turn would broadcast . . . Nicola Tesla*: Wait, "Sanguine Concept," 84–85.

58 *The ELF signal goes . . . into the ocean*: U.S. Navy, Navy Fact File, "Ex-

tremely Low Frequency Transmitter Site Clam Lake, Wisconsin," 3, online at http://enterprise.spawar.navy.mil/UploadedFiles/fs_clam_lake_elf2003.pdf.

58 *Besides nonattenuation, ELF signals ... carry little information*: Merrill, "Some Early," 361.

58 *Christofilos said this system ... cost $138 million*: Christofilos, "ELF—Communication System," 145.

58 *His intuition about physics ... than an analytical theorist*: Bromberg, *Fusion*, 122.

59 *He's buried in a Livermore ... Odd Fellows*: "Memory Gardens Odd Fellows Memorial Park," online at http://www.l-ags.org/cem_liv/mg21.html.

59 *Astron and Seesaw ... didn't die until Christofilos did*: On Astron, see Bromberg, *Fusion*, 202, 204. On Seesaw, see *Advanced Research Projects Agency, 1958–1974*, IX–32.

59 *Christofilos/Courant cosmotron, ... in, respectively, 1980 and 1988*: "Brookhaven and the Nobel Prize," online at http://www.bnl.gov/bnlweb/history/nobel/.

59 *and ARPA's budget for Jason had doubled to over $500,000*: Hal Lewis, interview by Finn Aaserud, AIP, July 6, 1986, 20;

59 *The ARPA historians said Jason's early years included its most significant contributions*: U.S. Department of Defense, *Advanced Research Projects Agency, 1958–1974*, IV–35.

59 *An unpublished in-house history ... productivity was "exceptional"*: MacDonald, "Early Years," 6.

60 *The ARPA historians said that while Jason's work on Defender ... decision to sign the Limited Test Ban Treaty*: U.S. Department of Defense, *Advanced Research Projects Agency, 1958–1974*, IV–34–35.

60 *A subsequent ARPA director ... "Defense Department has available"*: Charles Herzfeld, interview by Finn Aaserud, AIP, July 28, 1991, 11.

60 *"There was tugging and hauling ... we were very cocky"*: Sam Treiman, interview by Finn Aaserud, AIP, December 18, 1986, 25.

61 *"I found the work interesting ... and dazzle people"*: Ibid., 40–41.

61 *"could do something towards at least preventing a genuine holocaust"*: Edwin Salpeter, interview by Finn Aaserud, AIP, March 30, 1978, 61.

CHAPTER FOUR: **Heroes**

63 *"things were bent out of shape in Vietnam"*: Henry Kendall, interview by Finn Aaserud, AIP, November 26, 1986, 22.

64 *Jason briefly got into the social science . . . "and figure out a solution"*: Johnson, *Strange Beauty*, 186, 256–57.

64 *"had little direct impact . . . done on the subject in 1965"*: Nierenberg, "DCPG," 2.

65 *They were calling themselves the Cambridge Discussion Group . . . and George Kistiakowsky*: Christopher Twomey, "McNamara Line," 7.

66 *"were impressive . . . group was up to speed"*: Nierenberg, "DCPG," 3.

67 *"Dear Mr. Secretary . . . George Kistiakowsky for Jason-East"*: George B. Kistiakowsky to Robert McNamara, June 23, 1966. U.S. Department of State, *Foreign Relations of the United States, 1964–1968*, vol. 4, *Vietnam 1966*, online at http://www.state.gov/www/about_state/history/vol_iv/154_174.html.

67 *The second study . . . "was essentially nil"*: Vitale, *War Physicists*, 159.

67 *Kistiakowsky's letter had suggested "selective electronic jamming" of the enemy's communications*: Kistiakowsky to McNamara, June 23, 1966.

68 *"did not mince words or fudge its conclusions" . . . "the tenacity and recuperative capabilities of the North Vietnamese"*: Pentagon Papers, 4:115–116, 119.

69 *Two months later, in October 1966, . . . waiting us out, McNamara concluded*: Robert McNamara to Lyndon Johnson, memorandum, Washington, October 14, 1966. U.S. Department of State: *Foreign Relations of the United States, 1964–1968*, vol. 4, *Vietnam 1966*, online at http://www.state.gov/www/about_state/history/vol_iv/253_271.html.

69 *"probably the most categorical rejection" . . . edited slightly for effectiveness*: Pentagon Papers, 4:222–23.

70 *McNamara wrote that of all the studies. . . . quoted that famous first sentence*: McNamara, Blight, and Brigham, *Argument Without End*, 341.

70 *In his farewell speech . . . "the previous fall by the Jasons"*: Pentagon Papers, 4:231.

70 *In January 1966, while Jason . . . who would have to carry it out*: Pentagon Papers, 4:112–14.

70 *Within a few months, by March 1966, . . . "flexibility in employment of forces"*: MACV, *Command History*, 3:1071. Pentagon Papers, 4:114.

71 *Meanwhile the Cambridge Discussion Group . . . "defoliation techniques, and area-denial weapons"*: Pentagon Papers, 4:114–15.

71 *Cambridge stayed interested, though . . . troops to construct or guard it*: Nierenberg, "DCPG" 4.

72 *Garwin, who was a brand-new Jason . . . "wise old man for them"*: Richard Garwin, interview by Finn Aaserud, AIP, June 24, 1991, 18, 20.

72 *The briefings on the trail . . . "nodes, rest camps, bypasses, and so on"*: Nierenberg, "DCPG," 2.

73 *"We weren't allowed on the ground . . . done from the air"*: Richard Garwin, interview by Finn Aaserud, AIP, June 24, 1991, 19.

73 *"The Air-Supported Anti-Infiltration Barrier"*: All details of Jason's anti-infiltration barrier have come from U.S. Department of Defense, *Air-Supported Anti-Infiltration Barrier.*

75 *"Considerable cleverness . . . to taper off"*: Ibid., 1, 3, 7, 10, 13, 55.

76 *"Motives . . . pure as the driven snow"*: Hal Lewis, interview by Finn Aaserud, AIP, July 6, 1986, 25.

76 *Toward the end of the summer . . . a summary of the report*: Nierenberg, "DCPG," 5.

76 *On September 3 McNamara wrote . . . CINCPAC, with the same request*: Pentagon Papers, 4:123; MACV, *Command History,* 3:1072.

77 *Meanwhile the military had gotten . . . ordered the barrier into reality*: Pentagon Papers, 4:123–24; MACV, *Command History,* 3:1072–73.

77 *"unlimited acquisition freedom . . . traditional channels"*: Rego, "Anti-Infiltration Barrier Technology," 8.

77 *The deadline . . . one year away*: Ibid., 5.

78 *The whole barrier system was code-named Practice Nine*: MACV, *Command History,* 3:1073; Rego, "Anti-Infiltration Barrier Technology," 7.

78 *On October 14 McNamara . . . divert too many resources*: Pentagon Papers, 4:126–27, 130.

78 *Within a month the third . . . "concurred in a barrier undertaking"*: MACV, *Command History,* 3:1075.

78 *On January 12, 1967 . . . the "highest national priority"*: Rego, "Anti-Infiltration Barrier Technology," 8.

79 *For the air-supported barrier, . . . noisemakers and noise detectors be improved*: Ibid., 12.

80 *The slow, low-flying, two-engined . . . were probably never used*: Ibid., 31–32, 48. Chris Jeppeson, "Acoubuoy, Spikebuoy, Muscle Shoals and Igloo White 1999," online at http://home.att.net/~c.jeppeson/igloo_white.html. Larry Westin, "A Short History of the 553rd Reconnaissance Wing," June 7, 1999; updated, October 12, 2004; online at http://personalpages.tdstelme.net/~westin/batcat/cathst.txt.

80 *On June 13 DCPG changed the name . . . to Dye Marker:* MACV, *Command History,* 3:1082–84.

80 *About the same time another field test . . . it said they weren't:* "Air Force Hunting 'Mini-Mines' Lost Off the Beaches in Florida," *New York Times,* July 17, 1967, 35; " 'Mini-Mines' Found on Florida Beach Blind Airman," *New York Times,* July 18, 1967, 26.

80 *For this and other reasons, the highly classified . . . creative variations:* Dickson, *Electronic Battlefield,* 47–50.

81 *Taking control of the increasing publicity, . . . "barbed wire to highly sophisticated devices":* MACV, *Command History,* 3:1086.

81 *"I understand you were surprised" . . . better than they used to:* Robert McNamara (secretary of defense) to Lyndon Johnson, memorandum, September 11, 1967, Johnson Library, National Security File, Country File, Vietnam, 2D Barrier, Secret. Available online at U.S. Department of State, *Foreign Relations, 1964–1968,* vol. 5, *Vietnam 1967,* online at http://www.state.gov/r/pa/ho/frus/johnsonlb/v/13158.htm.

81 *At the same time DCPG changed . . . and Mud River, for the antitruck barrier:* MACV, *Command History,* 3:1086–87.

81 *By the end of November 1967 Mud River . . . went into operation:* Rego, "Anti-Infiltration Barrier Technology," 23–25.

81 *At first it wasn't working well . . . "confirmation for sake of confirmation":* Ibid., 26; MACV, *Command History,* 2:921.

82 *They put on earphones . . . " 'we lost four trucks' ":* Hal Lewis, interview by Finn Aaserud, AIP, July 6, 1986, 36.

82 *the first weeks of Mud River . . . "requesting data and explanations":* MACV, *Command History,* 2:921.

82 *the antitruck Mud River . . . to the siege at Khe Sanh:* Ibid., 2:921–22.

82 *Khe Sanh was a marine base . . . same North Vietnamese general:* Nalty, *Air Power,* 5–8, 24; "Confrontation at Khe Sanh," *New York Times,* February 18, 1968; U.S. Senate, *Hearings before the Electronic Battlefield Subcommittee,* 80–81.

83 *The number of marines was actually . . . it was operating:* Nalty, *Air Power,* iii; U.S. Senate, *Hearings before the Electronic Battlefield Subcommittee,* 83, 90; MACV, *Command History,* 2:922.

83 *A Marine captain and intelligence officer . . . "and we were ready for them":* Baig briefing Jason is from MacDonald, "Jason and DCPG," 10–11. Baig's letter is in U.S. Senate, *Hearings before the Electronic Battlefield Subcommittee,* 84–86.

84 *Throughout the long siege . . . beeped when a sensor triggered*: MACV, *Command History*, 2:922–23; MacDonald, "Jason and DCPG," 11–12.

85 *About two years later, in mid-November . . . sensors were similarly useful*: U.S. Senate, *Report of the Electronic Battlefield Subcommittee*, 1.

85 *"I have got to say that . . . have to say no, I wouldn't"*: U.S. Senate, *Hearings before the Electronic Battlefield Subcommittee*, 94–95.

86 *In April 1968, after the siege . . . "was a new ball game"*: Ibid., 3–4, 11–13.

86 *"Virtually every U.S. ground combat unit . . . detect the enemy"*: U.S. Senate, *Report of the Electronic Battlefield Subcommittee*, 6.

86 *The senators heard from all the services . . . "wants to be without them"*: Ibid., 6–10; House quote, 10.

86 *Major General Deane set out the sensors . . . "fiddle with them"*: U.S. Senate, *Hearings before the Electronic Battlefield Subcommittee*, 8.

86 *A year before the hearings, in October . . . "application of highly lethal firepower"*: Westmoreland, talk to annual luncheon, Association of the United States Army, Sheraton Park Hotel, Washington, D.C., October 14, 1969, online at www.stanford.edu/group/mmdd/SiliconValley/Westmore-land/westmoreland.rtf.

87 *During the hearings in 1970 Senator Barry Goldwater . . . "since gunpowder"*: U.S. Senate, *Hearings before the Electronic Battlefield Subcommittee*, 3.

87 *In the first place, the antitroop . . . intended for the antitroop barrier*: MACV, *Command History*, 2:925.

87 *But the military found these sensors . . . needed to find them*: Ibid., 2:923.

87 *the sensor system . . . used in Laos against trucks*: Ibid., 2: 919; Rego, "Anti-Infiltration Barrier Technology," 32–38.

87 *To assess success, the military kept data . . . attacking the South*: Ibid., 51.

88 *An air force historian summed up . . . "affect the outcome of the war"*: Ibid., 52.

88 *In January 1968, just as the sensors . . . resigned from DCPG's science advisory group*: Kevles, *Physicists*, 406.

88 *Charles Townes replaced Kistiakowsky . . . "simply didn't want this McNamara Wall"*: Townes, interview by Suzanne Riess, University of California, Berkeley.

89 *In June 1972 DCPG was disbanded*: Department of the Army, *Historical Summary, Fiscal Year 1972,* compiled by William Gardner Bell, Center of Military History, United States Army, Washington, D.C., 1974, online at http://www.army.mil/cmh-pg/books/DAHSUM/1972/ch10.htm.

89 *Henry Kendall, who had been, . . . "Jason experience deeply educating"*: Kendall, interview by Finn Aaserud, AIP, November 25–26, 1986, 22, 28.

89 *"a totally new technology that we now call smart weapons"*: Stephen Lukasik, interview by Finn Aaserud, AIP, April 21, 1987, 50.

CHAPTER FIVE: Villains

91 *"convey the feelings of the group, without threats, to people in the highest accessible places"*: Hal Lewis, Memorandum for Jason Members, March 29, 1968, in John A. Wheeler Papers, Series I, B:W564, American Philosophical Society, Philadelphia.

91 *"The subject is by no means closed"*: Ibid.

92 *"But I just felt so uncomfortable, I couldn't"*: Sam Treiman, interview by Finn Aaserud, AIP, December 18, 1986 51, 53.

92 *Henry Kendall, an early Jason . . . "way to change national policy"*: Kendall, interview by Finn Aaserud, AIP, November 25–26, 1986, 29.

92 *Dyson tried resigning . . . "have any such job in view"*: Hal Lewis to Freeman Dyson, February 19, 1968; Dyson to Lewis, March 18, 1968; both in Dyson's files.

92 *"We got taken to the cleaners"*: Marvin Goldberger, interview by Finn Aaserud, AIP, February 12, 1986, 5.

93 *"Many people who got involved . . . keep your guard up"*: Sidney Drell, interview by Finn Aaserud, AIP, July 1, 1986, 32.

93 *loose talk around the Pentagon . . . "would close that pass (and others) for good"*: "An Insider's Account: Seymour Deitchman," online at http://www.nautilus.org/VietnamFOIA/report/insider.html.

94 *The Mu Gia Pass was one . . . repaired and reused:* "Background: Targeting Ho Chi Minh Trail," online at http://www.nautilus.org/VietnamFOIA/background/HoChiMinhTrail.html.

94 *Passes like the Mu Gia . . . for tactical nuclear weapons:* "An Insider's Account," online at http://www.nautilus.org/VietnamFOIA/report/insider.html.

94 *"was a horrible idea"*: Steven Weinberg, interview by Finn Aaserud, AIP, June 28, 1991, 14.

94 *"would cast doubt on the impartiality . . . concentrated on purely military issues"*: Author's Commentaries, Steven Weinberg, online at http://www.nautilus.org/VietnamFOIA/report/JASONs.html#Weinberg.

95 *The Jasons' report covers . . . those weapons would be*: U.S. Department of Defense, *Tactical Nuclear Weapons in Southeast Asia*, 2.

95 *A few hundred tactical . . . "or between NVN and Laos"*: Ibid., 13.

95 *However, the Jasons figured . . . by fifty thousand men in a month or two*: Ibid., 14.

95 *"The main weakness of tree . . . blown down once"*: Ibid., 13.

96 *Using fewer bombs . . . would not be tactical but "anti-population"*: Ibid., 19.

96 *"And in spite of it all, the basic [North Vietnamese] system . . . bombardment would slacken"*: Ibid., 27.

97 *If an attack were coordinated . . . "would be essentially annihilated"*: Ibid., 6.

97 *The report seemed uncomfortable . . . "to carry this scenario further"*: Ibid., 40.

97 *Weinberg said the Jasons' analysis . . . "I would not have helped to write it"*: Steven Weinberg, Author's Commentaries, online at http://www.nautilus.org/ VietnamFOIA/report/JASONs.html#Weinberg.

98 *Weinberg said that he resigned . . . "was useful or not"*: Steven Weinberg, interview by Finn Aaserud, AIP, June 28, 1991, 20.

98 *The Joint Chiefs of Staff asked whether . . . make contingency plans*: Earle G. Wheeler (Chairman of the Joint Chiefs of Staff) to Lyndon Johnson, memorandum, February 3, 1968, note 5; online at http://www.state.gov/r/pa/ho/frus/johnsonlb/vi/13690.htm and http://www.mtholyoke.edu/acad/intrel/vietnam.htm.

98 *No nuclear weapons were currently . . . presumably nearby*: Tannenwald, *Nuclear Taboo*; "Nuclear Weapons and the Vietnam War," online at http://www.nautilus.org/VietnamFOIA/background/NuclearWeapons.html.

98 *On Monday, February 5 . . . indeed was the case*: John Finney, "Anonymous Call Set off Rumors of Nuclear Arms for VN," *New York Times*, February 12, 1968, 1–2; Michael Klare, "The Secret Thinkers," *Nation*, April 15, 1968, 503–504. Richard Garwin, interview by Finn Aaserud, AIP, June 24, 1991, 21.

99 *On February 11 Johnson . . . stop making contingency plans*: Wheeler to Johnson, memorandum, February 3, 1968, Note 5.

99 *Subsequently a graduate student . . . "someone who had participated in this study"*: Robert Gomer, Author's Commentaries, online at http://www. nautilus.org/VietnamFOIA/report/JASONs.html#GomerCom.

100 *According to the science historian Daniel Kevles, . . . 20 percent of Caltech's budget*: Kevles, *Physicists*, 402.

100 *Kevles cites numbers. . . "the emergence of radical student activism"*: Ibid., 405–406.

100 *In April 1970 an underground newspaper . . . stolen minutes of a Jason meeting*: Student Mobilization Committee to End the War in Vietnam, "Counterinsurgency Research on Campus EXPOSED," *Student Mobilizer* 3, no. 4, April 2, 1970, 3–16.

101 *"For three weeks . . . and American activities there"*: Deitchman, *Best-Laid Schemes*, 304–305.

101 *Sometime later a student broke . . . to the Student Mobilization Committee*: Eric Wolf and Joseph Jorgensen, "Anthropology on the Warpath in Thailand," *New York Review of Books*, November 19, 1970, 27.

101 *"We have never . . . maybe we can contribute to it"*: Student Mobilization Committee, "Counterinsurgency Research," 11.

101 *When Gell-Mann asked . . . "I want tools"*: Ibid., 8.

101 *Gell-Mann said . . . "other negatives have on villager attitudes?"*: Ibid., 10.

102 *Seven months later two anthropologists . . . intertwining was wrong*: Wolf and Jorgensen, "Anthropology," 27.

102 *The RAND analyst . . . all seven thousand pages of the history and leaked it*: Daniel Ellsberg, "Presidential Decision and Public Dissent," July 29, 1998. Interview by Harry Kreisler for Conversations with History, http:// globetrotter.berkeley.edu/people/Ellsberg/ellsberg98-0.html.

103 *"Coming as they did . . . influence on McNamara's thinking"*: *Pentagon Papers*, 4:111–12.

103 *In the barrier report . . . "the ineffective air war against North Vietnam."* Ibid., 4:122.

103 *At Columbia University forty-nine faculty members . . . from Columbia*: Henry Foley to William McGill, May 1, 1972, Henry Foley Papers, Box 16, Rare Book and Manuscript Library, Columbia University.

104 The Philadelphia Inquirer *ran an article . . . "moral crisis he felt*

from his role in Jason": Joel Shurkin, "The Secret War Over Bombing," *Philadelphia Inquirer*, February 4, 1973, 1.

105 *When he insisted on giving his lecture . . . "bodily expelled from the Institute*": Vitale, *War Physicists*, 112.

105 *A French poster . . . "the American war crimes.*": Ibid., 86.

105 *Italian physicists circulated what they called the "Trieste letter" . . . "military-industrial complex of the big powers*": Ibid., 115.

106 *"Imagine a discussion . . . and those who work on 'cyclon B'*": Ibid., 122.

106 *"If you sincerely want . . . but they have no effect on the war*": Ibid., 119.

106 *Schwartz had done his doctoral work . . . "if not overtake them*": Charles Schwartz, interview by Finn Aaserud, AIP, May 15, 1987, 9.

106 *Drell offered Schwartz . . . "You know, the spoiled child*": Ibid., 14.

106 *The following year he was made . . . " 'Get me a good job somewhere.' *": Ibid., 20.

106 *He ended up at Berkeley where, in 1962 . . . "science in the government*": Ibid., 27; Kenneth Watson, interview by Finn Aaserud, AIP, February 10, 1986, 9.

107 *Schwartz spent one summer . . . but was not asked to return*: Charles Schwartz, interview by Finn Aaserud, AIP, May 15, 1987, 9.

107 *A few years later his brother was killed . . . or SESPA*: Ibid., 36, 56–57.

107 *SESPA's initial announcement . . . "and political freedom*": Charles Schwartz, "A Skeletal Archive of Science for the People: Call for the First Meeting," online at http://socrates.berkeley.edu/%7Eschwrtz/SftP/Contents.html.

107 *"Now, within Jason . . . identified as war criminals*": Charles Schwartz, interview by Finn Aaserud, AIP, May 15, 1987, 69.

107 *In 1972 SESPA published* Science Against the People *. . . the Jasons talked openly*: SESPA, *Story of Jason*.

107 *they saw him as a colleague, Schwartz said*: Charles Schwartz, interview by Finn Aaserud, AIP, May 15, 1987, 73.

107 *Schwartz sent the small profiles . . . "needed to be held accountable*": Ibid., 71.

108 *Hal Lewis had written a letter . . . "a veil of 'scientific objectivity' and military secrecy*": SESPA, *Story of Jason*, 28–29.

108 *"Charlie Schwartz hates Jason . . . part of life*": Hal Lewis, interview by Finn Aaserud, AIP, July 6, 1986, 39.

108 *Every Wednesday SESPA's New York chapter . . . sometimes at the Jasons' homes*: Shapley, "Jason Division," 461–62.

109 *Beginning in March 1973 and continuing . . . a quiet delight in killing"*: Foley Papers, Jason/SESPA file, Box 16, Columbia University.

109 *One leaflet recounted an overheard story . . . "what's a few dead babies or mothers?"*: Deborah Wallace, "The Spirit of Jason," October 5, 1972: attachment to Richard Garwin to Bruce Smith, letter, November 2, 1972, in Foley Papers, Jason/SESPA file, Box 16, Columbia University.

109 *Garwin said the story wasn't true . . . "the Jews in Germany"*: Shapley, "Jason Division," 462.

109 *SESPA named Jasons . . . home addresses of three of them*: Foley Papers, Jason/SESPA file, Box 16, Columbia University.

109 *"You are the worst kind . . . the Nazi officers, were"*: Laurice Nassif to Henry Foley, August 9, 1972, in Foley Papers, Jason/SESPA file, Box 16, Columbia University.

109 *Leon Lederman . . . disarmament, and demilitarization*: Leon Lederman, "SESPA Kills, or Jason Fights Back," n.d., 1–3, in Foley Papers, Jason/SESPA file, Box 16, Columbia University.

110 *A group of maybe fifty people . . . stay where it was*: William McGill, talk given to B'nai Brith's Anti-Defamation League, November 16, 1972, 4–7, in Foley Papers, Jason/SESPA file, Box 16, Columbia University.

111 *Finally physicists outside Pupin . . . "horror" and "shame"*: John Darnton, "Lab Occupation Ends," *New York Times*, April 29, 1972, 1, 14.

111 *The next morning Columbia's administration . . . and the activists left*: McGill talk, in Foley Papers, 7.

112 *"should be tried for war crimes"*: Garwin and Baldwin, "Academic Freedom," 272, 276.

113 *"have caused terrible wounds among Vietnamese civilians"*: Vitale, *War Physicists*, 152.

113 *Jasons wrote back saying . . . fewer civilians would have died*: Ibid., 154–56.

113 *Another exception . . . but he didn't explain the delicacies*: E.H.S. Burhop, "Scientists and Soldiers: America's 'Jason Group' Looks Back on its Vietnam Involvement," *Bulletin of the Atomic Scientists*, November 1974, 8.

116 *Edwin Salpeter, an astrophysicist . . . dropped out of Jason*: Vitale, *War Physicists*, 52–53.

CHAPTER SIX: **Changes**

118 *A sample of headlines . . . "Scientists Urged to Improve Image"*: Sandra Blakeslee, "25 Form Council to Make Scientists Aware of Consequences of Their Work," *New York Times*, February 18, 1970, 11. Walter Sullivan, "Scientists Urged to Improve Image," *New York Times*, December 27, 1970, 33. "A Hippocratic Oath By Scientists Urged," *New York Times*, January 19, 1971, 62.

118 *During the Johnson administration . . . declined to reappoint PSAC*: Kevles, *Physicists*, 411–13. PSAC and the White House dining hall from Smith, *Advisers*, 168.

119 *ARPA's budget . . . off a cliff*: U.S. Department of Defense, *Advanced Research Projects Agency, 1958–1974*, fig I-1, I-2.

119 *Jason's "esprit de corps" was also in decline*: U.S. Department of Defense, *Advanced Research Projects Agency, 1958–1974*, IV–35.

119 *"In the earlier years . . . we couldn't always do wonders"*: Sam Treiman, interview by Finn Aaserud, AIP, December 18, 1986, 71.

119 *And Jason's studies—as usual . . . "Jason productivity" decreased significantly*: U.S. Department of Defense, *Advanced Research Projects Agency, 1958–1974*, IV–35.

121 *Though he had begun . . . "helping where I could"*: Stephen Lukasik, interview by Finn Aaserud, AIP, April 21, 1987, 7–8.

121 *Not only was ARPA a "new vibrant agency" . . . "what it should be spent on"*: Ibid., 12.

121 *"emphasized the authority of the director"*: U.S. Department of Defense, *Advanced Research Projects Agency, 1958–1974*, IX–17.

122 *In early 1967 Lewis . . . "they wanted access to places A, B, and C"*: Stephen Lukasik, interview by Finn Aaserud, AIP, April 21, 1987, 28.

124 *"I had to explain . . . And that began to shut up some of the critics"*: Charles Herzfeld, interview by Finn Aaserud, AIP, July 28, 1991, 10.

125 *"You have to warn them . . . an idiot because he cannot integrate a simple differential equation"*: Ibid., 15.

125 *"Strictly speaking . . . IDA ran the Jasons"*: Stephen Lukasik, interview by Finn Aaserud, AIP, April 21, 1987, 33.

127 *"The Jasons became absolutely useful . . . well coupled-in"*: Ibid., 40.

129 *"I am an oddity . . . observational and natural sciences"*: Walter Munk, interview by Finn Aaserud, AIP, June 30, 1986, 11.

131 *"an example of a phenomenon . . . killed by stubborn facts"*: Mitre, Jason, "Neutrino Detection Primer," 3–2.

133 *"now we know less and less . . . perfidious trend"*: Sidney Drell, interview by Finn Aaserud, AIP, July 1, 1986, 28.

134 *"Jasons wanted to work on things . . . rather than war-related projects"*: Gordon MacDonald, interview by Finn Aaserud, AIP, April 16, 1986, 16.

140 *ARM began taking measurements . . . climatic conditions*: "About ARM," online at http://www.arm.gov/about/.

144 *Treiman noticed . . . "the age was more establishmentarian"*: Sam Treiman, interview by Finn Aaserud, AIP, December 18, 1986, 56–57.

CHAPTER SEVEN: **Matching**

149 *The focus of the U.S. worry was the air force's new missile . . . so they'd be hard to target*: Richard Garwin, review of Frances FitzGerald, *Way Out There in the Blue*, in *Los Angeles Times Book Review*, April 30, 2000, online at http://www.fas.org/rlg/000400-fire.htm. Federation of American Scientists, "LGM-118A."

149 *The basing systems . . . into a small area in Wyoming"*: Dyson, *Weapons*, 63.

150 *partly because the Trident missiles were thirty-four feet high*: U.S. Navy fact file, "*Trident* Fleet Ballistic Missiles" online at http://www.chinfo. navy.mil/navpalib/factfile/missiles/wep-d5.html.

150 *The air force's MX missile was two times bigger, seventy-one feet high*: Federation of American Scientists, "LGM-118A."

150 *SUMs would be mobile, fifty to one hundred of them in the shallow waters along the continental coasts*: "Navy Funds Trident Replacement Study," *Aviation Week & Space Technology*, April 30, 1979, 193. George Wilson, "Pentagon Consultant Urges Small Subs to Launch New Missiles," *Washington Post*, February 7, 1979, A8.

151 *SUMs would have small crews of around twelve . . . closer to ten thousand tons*: Sidney Drell and Richard Garwin, "Basing the MX Missile: A Better Idea," *Technology Review*, May 1981, 20–29. Daniel S. Greenberg, "Missiles at Sea," *Washington Post*, August 7, 1979, A19. Richard Garwin, "From PSAC to 'Stockpile Stewardship,' " talk given at Sid Drell Symposium, July 31, 1998, online at http://www.slac.stanford.edu/gen/meeting/ssi/1998/Drell.

151 *A defense scientist said that a "Soviet barrage" . . . crush the*

SUMs: Donald M. Snow, "The MX-Basing Mode Muddle," *Air University Review,* July–August 1980, online at http://www.airpower.maxwell.af.mil/airchronicles/aureview/1980/jul-aug/snow.html.

151 *A Pentagon official said . . . order them from Germany*: Clarence A. Robinson, "MX Basing Study Draws Fire From Congress," *Aviation Week & Space Technology,* March 23, 1981, 16.

151 *Congress's Office of Technology Assessment . . . in favor of SUMs*: Office of Technology Assessment, *MX Missile Basing.*

151 Aviation Week & Space Technology *said SUM stood for "shallow underwater mobile"*: "Navy Funds Trident Replacement Study."

155 *One way to remove distortion . . . will look the way it should*: Horace Babcock, "Adaptive Optics Revisited," *Science* 249, July 20, 1990, 253–57.

156 *They published their idea . . . astronomers most want to see*: R. A. Muller and A. Buffington, "Real-Time Correction of Atmospherically Degraded Telescope Images Through Image Sharpening," *Journal of the Optical Society of America* 64 (July 1974), 1200–10.

156 *That same summer . . . Dyson also published his work*: Freeman Dyson, "Photon Noise and Atmospheric Noise in Active Optical Systems," *Journal of the Optical Society of America* 65 (March 1975), 551–58.

156 *In addition to funding Jason, DARPA . . . and look at the faint thing*: Rettig Benedict, James Breckinridge, and David Fried, "Atmospheric-Compensation Technology Introduction," *Journal of the Optical Society of America* 11, no. 1 (January 1994), 257–60.

158 *To learn more, DARPA . . . the sodium guide star did the same*: Ibid., 259.

159 *Reagan's science advisory . . . heard of it before*: Drell, *Shadow of the Bomb,* 264; York, *Arms and the Physicist,* 246–48.

160 *The tightness of the security . . . sodium laser guide stars*: R. Foy and A. Labeyrie, "Feasibility of Adaptive Telescope with Laser Probe," *Astronomy and Astrophysics* 152, no. 2 (November 1985), L29–L31.

164 *Three years later, in 1994 . . . two of Jason's three reports*: W. Happer, G. J. MacDonald, C. E. Max, and F. J. Dyson, "Atmospheric-Turbulence Compensation by Resonant Optical Backscattering from the Sodium Layer in the Upper Atmosphere," *Journal of the Optical Society of America* 11, no. 1 (January 1994), 263–76.

168 *The following February SDI's director . . . "a stupid idea"*: David Perlman, "Basic System Flawed: Critical Secret Report on Star Wars," *San*

Francisco Chronicle, February 19, 1990, A2. Happer quote about fixing problems in R. Jeffrey Smith, "Pentagon Increases SDI Push; Public, Hill Support Eroding as Fear of Soviet Threat Fades," *Washington Post,* February 18, 1990, A1. Happer quote about program not being all a crock and lasers being stupid in "Board Responded to a Narrow Question: Claimed Endorsement by Physicists Disputed by Chairman," *Washington Post,* February 18, 1990, A18.

CHAPTER EIGHT: **Blue Collars, White Collars**

172   *One of the science-related issues . . . SST would go ahead*: Smith, *Advisers,* 172–74; John Herbers, "Supersonic Plane Gets Triple Study," *New York Times,* February 8, 1969, 62; Richard Garwin, interview by Finn Aaserud, AIP, June 8, 1967, 15. Office of Science and Technology, *Final Report of the Ad Hoc Supersonic Transport Review Committee,* March 30, 1969. Richard Garwin, "The SST Pressure," speech to Illinois Mathematics and Science Academy, Aurora, Ill., May 4, 1993, online at http://www.fas.org/rlg/930504-imsa.htm.

173   *Meanwhile, the Garwin report . . . Congress killed the SST*: "Studies Ordered On Fate of SST," *Washington Post,* February 8, 1969, A2. Robert Phelps, "Nixon Panel Expected to Urge Shelving of Supersonic Plane," *New York Times,* March 16, 1969, 1. "SST Recommendations Submitted to President," *Washington Post,* March 29, 1969, A3. "Report on SST Said to be Kept Secret," *New York Times,* March 13, 1970, 50. House Subcommittee on Department of Transportation and Related Agencies Appropriations, *Hearings on Supersonic Transport Program,* 70-H181-32, 91st Cong., 2nd sess., April 23, 1970, 980–94; Garwin quote about 50 747's, Ibid., 982; Garwin quote about immediate termination, Ibid., 988. Joint Economic Committee, Subcommittee on Economy in Government, Hearings on Supersonic Transport Development, 91st Cong., 2nd sess., May 7, 1970, May 11–12, 1970, pp. 904–20. Senate, Subcommittee on Department of Transportation and Related Agencies Appropriations, Hearings on 70-S181-34, 91st Cong., 2nd sess., August 28, 1970, 1621–1681. Weart, "Public and Climate Change."

173   *The Nixon administration was unusually convinced . . . everyone cites him*: Smith, *Advisers,* 172–75; "science bastards" quote, p. 175. David Z. Robinson, "Politics in the Science Advising Process," in Golden, *Science Advice,* 224–25; and James S. Coleman, "The Life, Death, and Potential Future of PSAC," in ibid., 197–99.

174 *In 1991, for example, Will Happer . . . and was fired*: Irwin Goodwin, "Happer Leaves DOE Under Ozone Cloud for Violating Political Correctness," *Physics Today* 46, no. 6 (June 1993), 89–91.

175 *And Garwin said repeatedly in interviews . . . "the matter would have stopped there"*: Richard L. Garwin, "Presidential Science Advising," in Golden, *Science Advice,* 177–90. Garwin quote about honesty of administration in ibid., 189.

175 *"perhaps it would have been better . . . but as a citizen"*: Garwin quotes about resignation, and opinions about newsworthiness and PSAC's integrity, in Richard Garwin, interview by Finn Aaserud, AIP, June 8, 1987, 18. Garwin quote about the right track and acting as a citizen in Golden, *Science Advice,* 190.

175 *Other science advisers to President Nixon . . . Nixon himself said no*: Gordon MacDonald testimony, Joint Economic Committee, Subcommittee on Economy in Government, Hearings on 70-J841-15, 91st Congress, 2nd sess., May 12, 1970, 1005–1007. The second PSAC member was also a former science adviser to the president, Jerome Wiesner, in Golden, *Science Advise,* 189.

176 *"I am traveling . . . 'orbiting' has only one 't'"*: Richard Garwin, e-mail to author, November 14, 2003.

180 *"I'm going to do a straight . . . to me that's Jason work"*: Sidney Drell, interview by Finn Aaserud, AIP, July 1, 1986, 20.

182 *By June 1995 the Department of Defense . . . test new bombs*: R. Jeffrey Smith, "Administration Debates Pentagon Proposal to Resume Nuclear Tests," *Washington Post,* June 18, 1995, A17. William J. Broad, "Atom Powers Want to Test Despite Treaty," *New York Times,* March 29, 1995, A6.

182 *They based the report . . . yields less than four pounds of TNT*: Mitre, Jason, "Nuclear Testing," 1, 6. A thorough and accessible write-up of the report is in Drell, "Physics and U.S. National Security."

184 *The Senate had been arguing . . . and had read closely*: Senate, *Discussion of the National Defense Authorization Act for Fiscal Year 1996,* 104th Cong., 1st sess., August 4, 1995, *Congressional Record* 141, no. 129, S11353–72.

184 *The report's summary and conclusions were read into the* Congressional Record: Ibid., S11368–69.

184 *The senators voted . . . "Tests Backed by Senate Are Unnecessary"*: R. Jeffrey Smith, "Physicists Say Small Nuclear Tests Backed by Senate Are Unnecessary," *Washington Post,* August 9, 1995, p. A20.

186 *But Jason had been careful . . . "an experienced cadre of capable scientists and engineers"*: Mitre, Jason, "Nuclear Testing," 3.

186 *A week after the Jason report . . . 148 countries had signed*: Federation of American Scientists, "Comprehensive Test Ban Treaty Chronology," online at http://www.fas.org/nuke/control/ctbt/chron.htm.

186 *Yet in the fall of 1999 . . . and professional organizations*: Senate Committee on Foreign Relations, *Final Review of the Comprehensive Nuclear Test Ban Treaty*, 106th Cong., 1st sess., October 7, 1999, 3–6, 34.

186 *Garwin and Drell both testified . . . to no avail*: Garwin in ibid., 112–33. Sidney Drell testimony, Senate Armed Services Committee, *Hearings on Stockpile Stewardship and the Comprehensive Test Ban Treaty*, October 7, 1999, online at http://www.fas.org/nuke/control/ctbt/conghearings/drell.pdf.

187 *Garwin gave a conference talk . . . "pleased to be an author of this document"*: Richard Garwin, "The Maintenance of Nuclear Weapon Stockpiles Without Nuclear Explosion Testing," talk to Nuclear Forces in Europe, 24th Pugwash Workshop on Nuclear Forces, September 26, 1995, online at http://www.fas.org/rlg/u242pugw.txt.

188 *NIF has been billed at various times . . . way out of line*: Hugh Gusterston, "NIFty Exercise Machine," *Bulletin of the Atomic Scientists* 51 (September–October 1995), 22–29; Jim Dawson, "NIF Moves Forward Amid Controversy," *Physics Today* 54, issue 1 (January 2001), 21, online at http://www.physicstoday.org/pt/vol-54/iss-1/p21.html.

188 *Jason's reports present . . . attracting scientists to the national labs*: Mitre, Jason, "Inertial Confinement," 1–5.

192 *"I understood that technical intelligence . . . set my whole course"*: Sidney Drell, interview by Finn Aaserud, AIP, July 1, 1986, 12–13.

194 *In 2002 Drell, a colleague, and Robert Peurifoy . . . to destroy underground bunkers*: Sidney Drell, Raymond Jeanloz, and Bob Peurifoy, "Bunkers, Bombs, Radiation," *Los Angeles Times*, March 17, 2002.

CHAPTER NINE: **Whither Jason?**

195 *Given the times . . . Sid Drell said it was "simply fortuitous"*: Drell, "Physics and National Security," S468.

195 *In 2004 the Union of Concerned Scientists . . . science advice to the government called "Flying Blind"*: Union of Concerned Scientists, "Scientific Integrity in Policymaking: An Investigation into the Bush Administration's

Misuse of Science," March 2004, online at http://www.ucsusa.org/global_
environment/rsi/index.cfm. Henry Kelly et al., "Flying Blind: The Rise, Fall,
and Possible Resurrection of Science Policy Advice in the United States,"
Federation of American Scientists, Occasional Paper no. 2, December 2004,
online at http://www.fas.org/static/pubs.jsp.

198 *Confusingly, somewhere in the process . . . had been acting on Rumsfeld's orders*: Rik Kirkland and Bill Powell, "Don Rumsfeld Talks Guns and
Butter," *Fortune* 146, no. 10 (November 18, 2002), 143–44.

199 The Chronicle of Higher Education *cited the Jason story . . . "the unflinching advice that government officials don't always want to hear"*: Ron
Southwick, "Elite Panel of Academics Wins Fight to Continue Advising Military," *Chronicle of Higher Education* 48, no. 39 (June 7, 2002), A26–A27. "The
Case of the Jasons," *Nature* 416, no. 6879 (March 28, 2002), 351.

201 *"Defense Department tasks . . . terribly many biology topics"*: Henry
Kendall, interview by Finn Aaserud, AIP, November 25–26, 1986, 26.

201 *Biological warfare by one country against another . . . it recommended additions to the system*: Block, "Living Nightmares."

202 *In preparing for that report, Jason had asked the secretary of the
navy . . . "is undesirable"*: Richard Danzig, remarks to Jason spring meeting, April 23, 1999, online at http://www.chinfo.navy.mil/navpalib/people/
secnav/danzig/speeches/jasn0423.txt.

202 *"Biodetection Architectures" . . . would change as the threat changed*:
Mitre, Jason, "Biodetection," 3, 7.

205 *"Biofutures," . . . allow biologists to predict the cell's behavior?*:
Mitre, Jason, "Biofutures."

207 *Jason physicists' interest in biology . . . was called "Harness the
Hubris"*: "Pursuing Arrogant Simplicities," *Nature* 416 (March 21, 2002),
247. V. Adrian Parsegian, "Harness the Hubris: Useful Things Physicists
Could Do in Biology," *Physics Today,* July 1997, 23–27. Also see Jonathan
Knight, "Bridging the Culture Gap," *Nature* 419, September 19, 2002,
244–46; and James Glanz, "Physicists Advance into Biology," *Science* 272,
May 3, 1996, 646–48.

209 *IDA's Defense Science Studies Group . . . including Callan, Happer,
and Koonin*: The Defense Science Studies Group, which calls itself DSSG,
sounds a little like a sibling of Jason, at least in its parentage—IDA and
DARPA—and in its broad education in the technical problems of the Defense
Department. The biggest difference is that DSSG members rotate out after two
years and so can develop neither the internal cohesiveness nor the relationship

with the sponsor that Jason has. A reasonable number of Jasons were recruited from DSSG. DSSG has a nice website at http://dssg.ida.org/index.html.

214 *DARPA has two Jason-like groups of advisers . . . focuses on materials science*: The group that focuses on information technology is Information Science and Technology, called ISAT. The other, which focuses on materials science, is the Defense Sciences Research Council, or DSRC. The membership of both groups is appointed by agreement between the group and DARPA. ISAT members are usually rotated out after three years; they spend about a week every summer working on studies. DSRC's members, who are often their field's luminaries, nominally rotate out, though some seem to stay on for years. DSRC meets for three weeks in the summer and is less likely to study specific defense problems than to review present and proposed defense programs. DSRC was born as the Materials Research Council and was nicknamed the Masons. Frank Fernandez, an ex-director of DARPA, says that both ISAT and DSRC have members whose research is supported by DARPA, but he is emphatic that DARPA's sole sponsorship of both groups does not compromise their independence, since the only purpose of both is "to tell DARPA that the emperor has no clothes."

229 *John Wheeler and Gordon MacDonald . . . in other words, irrevocable*: Gordon MacDonald, interview with Finn Aaserud, AIP, April 16, 1986, 27; John A. Wheeler, interview with Finn Aaserud, AIP, November 28, 1988, 73–79.

229 *Pure scientists in Europe . . . employed by it*: Bruce Smith, *The Advisers*, 7–9.

EPILOGUE: **Outcomes and Updates**

232 *In 2000, at age ninety-four, Bethe wrote a letter . . . missile defense tests*: "Nobel Laureates Warn Against Missile Defense Deployment," online at http://www.fas.org/press/000706-news.htm.

232 *in 2001 he wrote a letter with the same intent to the Congress*: Federation of American Scientists to congressional leader, November 12, 2001, online at http://www.fas.org/nobel.pdf.

232 *The present board of sponsors of the Federation of American Scientists . . . Richard Garwin and Steven Weinberg*: "Board of Sponsors," Federation of American Scientists, online at http://www.fas.org/static/board.jsp.

233 *"the oldest organization . . . weapons for any purpose"*: "About FAS," online at http://www.fas.org/static/about.jsp.

236 *"Star Wars II: Can We Be Protected by a Pig in a Poke?"*: Los Angeles *Times,* June 2, 2000, 9.

236 *Sanguine was to transmit at a frequency . . . had the proper conduct-ing properties*: Sanguine's frequency and antenna length are from Wait, "Sanguine Concept," p. 84. Wisconsin location is from U.S. Navy, "Ex-tremely Low Frequency," 3.

236 *Sanguine evolved into Seafarer . . . electromagnetic fields this strong*: Ibid., 1–7. Size and rate of message are from Senator Russ Feingold, speech to Senate on S.47, Project ELF Termination Act, 108th Cong., 1st sess., *Con-gressional Record* 149 (January 7, 2003).

236 *The studies are exhaustive . . . earthworms, and monkeys, among others*: Executive Summary, "An Evaluation of the U.S. Navy's Extremely Low Frequency Communications System Ecological Monitoring Program," National Research Council, National Academy Press, 1997, online at http://books.nap.edu/html/elf/.

237 *As of 2001, a U.S. senator from Wisconsin . . . to have ELF killed*: The cost of ELF comes from a speech by Senator Russ Feingold, online at http://www.senate.gov/~feingold/issuearea/defense.html. The 2003 rein-troduction is from Senator Russ Feingold, speech to Senate on S.47, Project ELF Termination Act, 108th Cong., 1st sess., *Congressional Record* 149, (January 7, 2003).

237 *In September 2004 the navy shut ELF down . . . "and the Cold War is over"*: U.S. Fleet Forces Command, press release 20-04, September 17, 2004. Philip Coyle (former assistant secretary of defense), interview, *Morning Edition,* National Public Radio, September 29, 2004, transcript.

240 *"Tactical Nuclear Weapons in Southeast Asia" . . . "frightening today as they were in 1967"*: The Nautilus Institute has an easily navigable website at http://www.nautilus.org/VietnamFOIA/background/background.html. The quote from Senator Dianne Feinstein is at http://www.nautilus.org/VietnamFOIA/Feinstein.html.

241 *The Electronic Barrier on the Mexican Border*: Senate, *Hearings be-fore the Electronic Battlefield Subcommittee,* 14. Ken Clawson, "U.S. Testing Sensors Along Mexican Border," *Washington Post,* July 18, 1970, A1. David Andelman, "U.S. Implanting an Electronic 'Fence' to Shut Mexican Border to Smuggling," *New York Times,* July 14, 1973, 1.

242 *Charles Schwartz was awarded tenure . . . abandoned the require-ment*: Charles Schwartz, interview by Finn Aaserud, AIP, May 15, 1987, 85, 96.

242 *Later, he stopped teaching . . . bodies to the defense contractors*: Ibid., 105.

242 *Eventually he stopped receiving raises . . . use their tenure to be activists*: Ibid., 96–97.

242 *"I feel very good about this kind of a life," he said*: Ibid., 110.

242 *Schwartz said he survived as a radical . . . "Most physicists pursue physics like that"*: Ibid., 83.

243 *In 1987 Schwartz told National Public Radio . . . "just keep the Pentagon more efficient"*: National Public Radio, December 6, 1987.

243 *The center's projects are two new kinds . . . each aimed in slightly different directions*: Center for Adaptive Optics at the Lick Observatory, online at http://cfao.ucolick.org/research/exao.php and http://cfao.ucolick.org/research/elts.php.

244 *Astronomy websites show adaptive optics . . . its tiny moon*: Neptune images at Center for Adaptive Optics at the Lick Observatory, online at http://cfao.ucolick.org/pgallery/io.php. Images of stars around black hole at the Canada-France-Hawaii Telescope at the University of Hawaii may be found at http://www.cfht.hawaii.edu/Instruments/Imaging/AOB/best_pictures.html#galactic%20center. Asteroid images may be found at the Southwest Research Institute website, http://www.boulder.swri.edu/~merline/press_release.

245 *The military is using adaptive optics . . . hundreds of miles away*: Louis A. Arana-Barradas, "Killer Laser's Scud Eraser," *Airman*, June 2000, online at http://www.af.mil/news/airman/0600/abl.htm. Sandra Erwin, "Air Force Mulls Expanding Role for Missile-Killer Lasers," *National Defense*, July 2000, online at http://www.nationaldefensemagazine.org/issues/2000/Jul/ Air_Force_Mulls.htm.

245 *the UCSC center is using adaptive optics . . . treat retinal disease*: Corinna Wu, "Supernormal Vision: A Focus on Adaptive Optics Improves Images of the Eye and Boosts Vision," *Science News*, November 15, 1997, online at http://www.sciencenews.org/sn_arc97/11_15_97/bob1.htm.

246 *In 2000 the U.S. General Accounting Office's . . . $1.8 billion over budget*: United States General Accounting Office, Report to the Subcommittee on Military Procurement, Committee on Armed Services, House of Representatives. *National Ignition Facility: Management and Oversight Failures Caused Major Cost Overruns and Schedule Delays*, August 2000, 1–9.

247 *By the end of 2004, Congress . . . outlook for NIF's success*: 108th Cong., 2nd sess., Report 108–792, *Making Appropriations for Foreign Oper-*

ations, Export Financing, and Related Programs for the Fiscal Year Ending September 30, 2005, and for Other Purposes, Conference Report to Accompany H.R. 4818, November 20, 2004, 952.

247 *At the end of June 2005 . . . that would keep NIF on track:* Mitre, Jason, "NIF Ignition," June 29, 2005, 3, 4–5.

247 *in early July, the Senate voted to stop funding its construction:* William Broad, "Senate Votes to Shut Down Laser Meant for Fusion Study," *New York Times,* July 2, 2005, online at http://www.nytimes.com/2005/07/02/politics/02laser.html.

247 *But back in 2003 . . . Congress added $75,000,000 to the program's budget:* Mitre, Jason, "Requirements for ASCI," October 2003, 69; 109th Cong., 1st sess., Senate Report 109-84, 109 S. Rpt. 84, Energy and Water Appropriations Bill, 2006 Report, June 16, 2005.

247 *In 2003 DARPA was deciding . . . "expert technical opinions":* Sharon Weinberger, "Scary Things Come in Small Packages," *Washington Post,* March 24, 2004, W15; Robert L. Park, "Hafnium: Congress Kills the Isomer Energy Release Program," *What's New,* June 4, 2004, online at http://www.aps.org/WN/WN04/wn041604.cfm; Senate Armed Services Committee, *Tactical Technologies,* Report 108–260, 108th Cong., 2nd sess.

248 *His sister, Alice . . . calculating things:* Schweber, *QED,* 476.

248 *Bethe eventually recommended . . . "best I have ever had or observed":* Ibid., 500.

248 *In 1948 Dyson drove from Cleveland . . . "blast and fire damage":* Dyson, *Disturbing the Universe,* 60–61.

249 *He says that doctoral students end up . . . "a badge of honor":* Dyson, *Eros to Gaia,* 195–96; Freeman Dyson, interview by Finn Aaserud, AIP, December 17, 1986, 6.

249 *He designed and forcefully backed . . . wrote a history of it, called* Project Orion: Dyson, *Project Orion.*

Sources

Interviews
[This list includes only those who agree to let their names be made public.]

Henry Abarbanel: San Diego, Calif., June 20, 2002

Paul Alivisatos: Berkeley, Calif., June 10, 2002

Steven Block: Stanford, Calif., June 6, 2002

Keith Brueckner: Kenneth Watson, and Marvin Goldberger, La Jolla, Calif., June 17, 2002

Curt Callan: Princeton, N.J., March 10, 2003

Ken Case: La Jolla, Calif., June 18, 2002

William Dally: Stanford, Calif., June 11, 2002

Seymour Deitchman: telephone, October 21, 2003

Sid Drell: Stanford, Calif., June 7 and 8, 2002; telephone, January 20, 2004

Freeman Dyson: Princeton, N.J., September 19, 1991, and March 10, 2003

Frank Fernandez: telephone, November 5 and 16, 2004

Val Fitch: Princeton, N.J., October 18, 2001

Alexander Flax: telephone, December 17, 2003

Ed Frieman: La Jolla, Calif., June 21, 2002; telephone, May 19 and 20, 2003, November, 17, 2003, January 23, 2004, and August 10, 2004

Robert Fugate: telephone, May 23, 2003

Richard Garwin: Members' Center, National Academy of Sciences, January 11, 2002, and April 6, 2002; telephone, November 17, 2003

Marvin Goldberger: San Diego, Calif., July 22, 1999; La Jolla, Calif., June 16 and 20, 2002; telephone, August 16, 2002, August 13, 2003, September 25, 2003, and October 13, 2003

Mildred Goldberger: La Jolla, Calif., July 22, 1999; telephone, June 17, 2002, and September 25, 2003

Robert Gomer: telephone, October 21, 2003

William Happer: Princeton, N.J., November 16, 2000, October 19, 2001, and December 13, 2002
Peter Hayes: telephone, November 13, 2003
Robert Henderson: McLean, Va., September 26, 2002, October 3, 2002
Paul Horowitz: Cambridge, Mass., December 20, 2001, February 15, 2002
Robert Kargon: Baltimore, Md., October 10, 2001
Randy Katz: Berkeley, Calif., June 4, 2002
Walter Kohn: Baltimore, Md., May 10, 2000
Steven Koonin: telephone, March 22, 2002; Pasadena, Calif., June 13, 2002
Hal Lewis: telephone, April 24, 2003
Stephen Lukasik: telephone, November 12 and 13, 2003; Arlington, Va., June 25, 2005
Gordon MacDonald: Cambridge, Mass., July 17, 2000, December 19 and 20, 2001, and February 13, 2002
Claire Max: Rosslyn, Va., April 26, 2003
Richard Muller: Berkeley, Calif., June 5, 2002
Walter Munk: La Jolla, Calif., June 19, 2002
David Nelson: Baltimore, Md., January 14, 2003
Aristides Patrinos: Washington, D.C., June 13, 2003
William Perry: telephone, October 29, 1991; Washington, D.C., October 7, 2002
William Press: telephone, October 23, 2003; Los Alamos, N.M., January 23, 2003; Arlington, Va., April 30, 2003
Lawrence Principe: Baltimore, Md., June 5, 2003
Burton Richter: Baltimore, Md., April 24, 2001
Jack Ruina: Cambridge, Mass., December 20, 2001, February 12, 2002; telephone, March 6, 2005
Roy Schwitters: Rosslyn, Va., November 23, 2003
William Thompson and Robert Duffner: telephone, May 29, 2003
Charles Townes: Berkeley, Calif., June 3 and 4, 2002
Wayne van Citters: telephone, January 30, 2004
Kenneth Watson: telephone, August 13, 2003
Steven Weinberg: telephone, October 22, 2003
Peter Weinberger: New York, June 10, 2003
John Wheeler: telephone, August 7, 1991; High Island, Me., July 3, 2001
M. Gordon Wolman: Baltimore, Md., July 18, 2001, February 6, 2002
Carl Wunsch: Baltimore, Md., October 21, 2002
Herbert York: La Jolla, Calif., June 16, 2002

Books and Articles

Aaserud, Finn. "Sputnik and 'the Princeton Three:' The National Security Laboratory That Was Not to Be." *Historical Studies in the Physical and Biological Sciences* 25, pt. 2 (1995), 185–239.

Badash, Lawrence, Joseph O. Hirschfelder, and Herbert Broida. *Reminiscences of Los Alamos, 1943–1945.* Dordrecht, Netherlands: D. Reidel Publishing Co., 1980.

Bethe, Hans. "Observations on the Development of the H-Bomb." In Herbert York, *The Advisors, Oppenheimer, Teller, and the Superbomb.* Stanford, Calif.: Stanford University Press, 1989.

———. *The Road from Los Alamos.* New York: AIP Press, 1991.

Block, Steven M. "Living Nightmares: Biological Threats Enabled by Molecular Biology," in Sidney Drell, Abraham D. Sofaer, and George D. Wilson, *The New Terror: Facing the Threat of Biological and Chemical Weapons.* Stanford, Calif.: Hoover Institution Press, 1999.

Boehm, George A. W. "The Pentagon and the Research Crisis." *Fortune* 57 (February 1958).

Bromberg, Joan Lisa. *Fusion: Science, Politics, and the Invention of a New Energy Source.* Cambridge: MIT Press, 1982.

Brueckner, Keith. Unpublished autobiography, 1986. Manuscript Biography Collection, Niels Bohr Library, American Institute of Physics, College Park, Maryland.

Buderi, Robert. *The Invention That Changed the World.* New York: Touchstone, 1997.

Burtless, Gary. *How Much Is Enough? Setting Pay for Presidential Appointees.* Brookings Institution, March 22, 2002, online at http://www.appointee.brookings.org/events/pay.pdf.

Charles Hard Townes. Interviews conducted by Suzanne B. Riess in 1991 and 1992. Regional Oral History Office, University of California, Berkeley, 1994. Available from the Online Archive of California, http://ark.cdlib.org/ark:/13030/kt3199n627.

Christofilos, Nicholas. "ELF—Communication System," *Ocean '72: IEEE International Conference in Engineering in the Ocean Environment,* IEEE Publication 72, September 1972, 145.

———. "The Argus Experiment." *Proceedings of the National Academy of Sciences* 45, no. 8 (August 15, 1959).

Deitchman, Seymour J. *The Best-Laid Schemes: A Tale of Social Research and Bureaucracy.* Cambridge: MIT Press, 1976.

Dickson, Paul. *The Electronic Battlefield*. Bloomington: Indiana University Press, 1976.

Drell, Sidney D. *Facing the Threat of Nuclear Weapons*. Seattle: University of Washington Press, 1983.

———. *In the Shadow of the Bomb: Physics and Arms Control*. New York: AIP Press, 1993.

———. "Physics and U.S. National Security," *Reviews of Modern Physics* 71, no. 2 (1999), S460–S470.

Dyson, Freeman. *Disturbing the Universe*. New York: Harper Colophon, 1979.

———. *Weapons and Hope*. New York: Harper and Row, 1984.

———. *Infinite in All Directions*. New York: Harper and Row, 1988.

———. *From Eros to Gaia*. New York: Pantheon, 1992.

———. *The Sun, the Genome, the Internet*. New York: Oxford University Press, 1999.

Dyson, George. *Project Orion: The True Story of the Atomic Spaceship*. New York: Henry Holt and Co., 2002.

Federation of American Scientists, "LGM-118A" (on Peacekeeper), online at http://www.fas.org/nuke/guide/usa/icbm/lgm-118.htm.

Fermi, Laura. *Atoms in the Family: My Life with Enrico Fermi*. Albuquerque: University of New Mexico Press, 1954.

Freiman, Edward. Interview by Naomi Oreskes and Ron Rainger, La Jolla, Calif., February 17, 2000. Transcript. Oral History Project of the H. John Heinz III Center for Science, Economics and the Environment. Online at http://www.heinzctr.org/Programs/SOCW/interviewers_colloquia.htm.

Frisch, Otto. *What Little I Remember*. Cambridge, Eng.: Cambridge University Press, 1979.

Garwin, Richard, and Frank Baldwin. "Academic Freedom, Research and the War." *Christianity and Crisis* 32, no. 21 (December 11, 1972).

Golden, William, ed. *Science Advice to the President*, 2nd ed. Washington, D.C.: AAAS Press, 1993.

Goodchild, Peter. *J. Robert Oppenheimer: Shatterer of Worlds*. New York: Fromm International Publishing Corp., 1995.

Gusterson, Hugh. "NIF-ty Exercise Machine." *Bulletin of the Atomic Scientists* 51 (September/October 1995), 22–29.

———. *Nuclear Rites: A Weapons Laboratory at the End of the Cold War*. Berkeley: University of California Press, 1996.

Hardy, John. "The Short, Eventful History of Adaptive Optics." In John Hardy, *Adaptive Optics for Astronomical Telescopes*. New York: Oxford University Press, 1998. Chap. 1.

Heilbron, J. L. *The Dilemmas of an Upright Man: Max Planck and the Fortunes of German Science*. Cambridge, Mass.: Harvard University Press, 1996.

Herken, Gregg. *Cardinal Choices: Presidential Science Advising from the Atomic Bomb to SDI*. New York: Oxford University Press, 1992.

Hoddeson, Lillian, et al. *Critical Assembly: A Technical History of Los Alamos During the Oppenheimer Years, 1943–1945*. Cambridge, Eng.: Cambridge University Press, 1993.

Johnson, George. *Strange Beauty: Murray Gell-Mann and the Revolution in Twentieth-Century Physics*. New York: Vintage Books, 1999.

Kevles, Daniel. *The Physicists: The History of a Scientific Community in Modern America*. New York: Vintage Books, 1979.

Kistiakowsky, George B. *A Scientist at the White House*. Cambridge, Mass.: Harvard University Press, 1976.

MacDonald, Gordon J. "Jason and DCPG—Ten Lessons." Talk, Jason's twenty-fifth Anniversary Celebration, November 30, 1984. Unpublished manuscript.

———. "Jason—The Early Years." Talk given to Jason, Washington, D.C., December 12, 1986. Unpublished manuscript.

MacDonald, Gordon J. F., and Charles A. Zraket. "Quality of Technical Decisions in the Federal Government." In William T. Golden, ed., *Science and Technology Advice to the President, Congress, and Judiciary*. New Brunswick, N.J.: Transaction Publishers, 1993.

McNamara, Robert S., James G. Blight, and Robert K. Brigham. *Argument Without End: In Search of Answers to the Vietnam Tragedy*. New York: Public Affairs, 1999.

Melissinos, A[drian]. C. "Nicholas C. Christofilos: His Contributions to Physics." Presentation to CERN Advanced Accelerator Physics Course, Rhodes, Greece, September 20–October 1, 1993. University of Rochester Department of Physics and Astronomy preprint.

Merrill, John. "Some Early Historical Aspects of Project Sanguine." *IEEE Transactions on Communications* 22, no. 4 (April 1974).

Mitre Corporation. Jason Program Office, "Neutrino Detection Primer," JSR-84-105, March 1988.

———. Jason, "Nuclear Testing, Summary and Conclusions." JSR-95-320, August 1995.

———. Jason, "Inertial Confinement Fusion Review," JSR-96-300, March 1996.

———. Jason, "Biodetection Architectures," JSR-02-330, February 2003.

———. Jason, "Biofutures," JSR-00-130, June 2001.

Nalty, Bernard C. *Air Power and the Fight for Khe Sanh*. Washington, D.C.: Office of Air Force History, U.S. Air Force, 1986.

"Navy Funds Trident Replacement Study." *Aviation Week & Space Technology*, April 30, 1979, 193.

Nautilus Institute for Security and Sustainability, " 'Essentially Annihilated': Nuclear First-Use in Vietnam: Implications for the War on Terror." Online at http://www.nautilus.org/VietnamFOIA/background/background. html.

Nierenberg, William. Interview by Naomi Oreskes and Ron Rainger, La Jolla, Calif., February 10, 2000. Transcript. Oral History Project of the H. John Heinz III Center for Science, Economics and the Environment. Online at http://www.heinzctr.org/Programs/SOCW/interviewers_colloquia.htm.

Nierenberg, William A. "DCPG—The Genesis of a Concept." *Journal of Defense Research*, Series B, Tactical Warfare (Fall 1969); declassified unpublished manuscript, November 18, 1971.

Oppenheimer, J. Robert. "Physics in the Contemporary World," in Martin Gardner, ed., *Great Essays in Science*. Amherst, N.Y.: Prometheus Books, 1944.

Pelliccia, Hayden. "Was Jason a Hero?" *New York Review of Books* (July 19, 2001), 53–56.

The Pentagon Papers: The Defense Department History of United States Decisionmaking on Vietnam, vol. 4, *Air War in North Vietnam, 1965–1968*. Boston: Beacon Press, 1971. Online at http://www.mtholyoke.edu/acad/intrel/pentagon/pent1.html.

Powers, Thomas. *Heisenberg's War: The Secret History of the German Bomb*. Cambridge, Mass., and New York: DaCapo Press, 2000.

Primack, Joel, and Frank von Hippel. *Advice and Dissent: Scientists in the Political Arena*. New York: Basic Books, 1974.

Rego, Robert. "Anti-Infiltration Barrier Technology and the Battle for Southeast Asia, 1966–1972." Thesis, Air Command and Staff College, Air University, Maxwell Air Force Base, Alabama, April 2000. Online at https://research.au.af.mil/papers/ay2000/acsc/00-147.pdf.

Rhodes, Richard. *The Making of the Atomic Bomb*. New York: Touchstone, 1988.

Richelson, Jeffrey T. *The Wizards of Langley: Inside the CIA's Directorate of Science and Technology*. Boulder, Colo.: Westview Press, 2002.

Schweber, Silvan S. *QED and the Men Who Made It: Dyson, Feynman, Schwinger, and Tomonaga*. Princeton, N.J.: Princeton University Press, 1994.

———. *In the Shadow of the Bomb: Bethe, Oppenheimer, and the Moral*

Responsibility of the Scientist. Princeton, N.J.: Princeton University Press, 2000.

Scientists and Engineers for Social and Political Action (SESPA). *Science Against the People: The Story of Jason*. Berkeley, Calif.: Berkeley SESPA, 1972.

Serber, Robert. *The Los Alamos Primer: The First Lectures on How to Build an Atomic Bomb*. Berkeley: University of California Press, 1992.

Shapley, Deborah. "Jason Division: Defense Consultants Who Are Also Professors Attacked," *Science* 179, February 2, 1973.

Shattuck, Roger. *Forbidden Knowledge: From Prometheus to Pornography*. New York: St. Martin's Press, 1996.

Smith, Alice Kimball. *A Peril and a Hope: The Scientists' Movement in America, 1945–47*. Chicago: University of Chicago Press, 1965.

Smith, Alice Kimball, and Charles Weiner. "Robert Oppenheimer: The Los Alamos Jeans." *Bulletin of the Atomic Scientists*, June 1980.

Smith, Bruce L. R. *The Advisers: Scientists in the Policy Process*, Washington, D.C.: Brookings Institution, 1992.

Tannenwald, Nina. *The Nuclear Taboo: The United States and the Nonuse of Nuclear Weapons Since 1945*. New York: Cambridge University Press, 2005.

Townes, Charles H. *How the Laser Happened: Adventures of a Scientist*. New York: Oxford University Press, 1999.

———. Interview by Suzanne Reiss, 1991, 1992. Regional Oral History Office, University of California, Berkeley. Online at http://ark.cdlib.org/ark:/13030/kt3199n62.

Traweek, Sharon. *Beamlines and Lifetimes: The World of High Energy Physicists*. Cambridge, Mass.: Harvard University Press, 1988.

Twomey, Christopher. "The McNamara Line and the Turning Point for Civilian Scientist-Advisors in Defence Policy." *Minerva* 37, no. 3 (Autumn 1999), 235–58. Online at http://www2.bc.edu/%7Etwomeych/Research.htm.

U.S. Atomic Energy Commission. *In the Matter of J. Robert Oppenheimer*, transcript of hearing before Personnel Security Board, Washington, D.C., April 12–May 6, 1954.

U.S. Congress. Office of Technology Assessment. *A History of the Department of Defense Federally Funded Research and Development Centers*. OTA-BP-ISS-157. Washington, D.C.: U.S. Government Printing Office, June 1995. Online at http://govinfo.library.unt.edu/ota/Ota_1/DATA/1995/9501.pdf.

———. *MX Missile Basing*. Washington, D.C.: U.S. Government Printing

Office, September 1981. Online at http://www.wws.princeton.edu/
~ota/disk3/1981/8116_n.html.

U.S. Department of Defense. *The Advanced Research Projects Agency,
1958–1974*. Washington, D.C.: Richard J. Barber Associates, December 1975.

U.S. Department of Defense. Advanced Research Projects Division. Institute for Defense Analyses. *Identification of Certain Current Defense Problems and Possible Means of Solution*. IDA-ARPA Study No. 1, August 1958. Defense Technical Information Center, Alexandria, Va.

U.S. Department of Defense. Institute for Defense Analyses. Jason Division. *Air-Supported Anti-Infiltration Barrier*. Study S-255, August 1966.

———. *Tactical Nuclear Weapons in Southeast Asia*. Study S-266, March 1967. Online at http://www.nautilus.org/VietnamFOIA/report/dyson 67.pdf.

U.S. Department of State. *Foreign Relations of the United States 1964–1968*, vol. 4, *Vietnam 1966*. Department of State Publication 10517, Office of the Historian, Bureau of Public Affairs. Washington, D.C.: U. S. Government Printing Office, 1998. Online at http://www.state.gov/www/about _state/history/vol_iv/index.html.

U.S. Congress. Senate. *Hearings before the Electronic Battlefield Subcommittee of the Preparedness Investigating Subcommittee of the Committee on Armed Services*. 91st Cong., 2nd sess., November 18, 19, 24, 1970. Washington, D.C.: U.S. Government Printing Office, 1970.

———. *Report of the Electronic Battlefield Subcommittee of the Preparedness Investigating Subcommittee of the Committee on Armed Services*, 92nd Cong., 1st sess. Washington, D.C.: U.S. Government Printing Office, 1971.

U.S. Military Assistance Command, Vietnam (MACV). *Command History*, vols. 2 and 3, 1967, 1968. Military History Branch, Office of the Secretary, Joint Staff, MACV.

U.S. Navy, Navy Fact File. "Extremely Low Frequency Transmitter Site Clam Lake, Wisconsin." Online at http://enterprise.spawar.navy.mil/ UploadedFiles/fs_clam_lake_elf2003.pdf.

Vitale, Bruno, ed. *The War Physicists: Documents About the European Protest Against the Physicists Working for the American Military Through the JASON Division of the Institute for Defence Analysis (IDA), 1972*. Naples, Italy: B. Vitale, 1976.

Wait, James R. "The Sanguine Concept." *Ocean '72: IEEE International Con-*

ference in Engineering in the Ocean Environment, IEEE Publication 72, September 1972.

Weart, Spencer. "The Public and Climate Change." In *The Discovery of Global Warming*. Cambridge, Mass.: Harvard University Press, 2003, online at http://www.aip.org/history/climate/Public.htm.

Wheeler, J. Craig. *The Krone Experiment*. New York: New American Library, 1986.

Wheeler, John Archibald, with Kenneth Ford. *Geons, Black Holes, and Quantum Foam: A Life in Physics*. New York: W.W. Norton and Co., 1998.

Wilson, Jane., ed. *All in Our Time: The Reminiscences of Twelve Nuclear Pioneers*. Chicago: Educational Foundation for Nuclear Science, 1975.

York, Herbert F. *Making Weapons, Talking Peace: A Physicist's Odyssey from Hiroshima to Geneva*. New York: Basic Books, 1987.

———. *Arms and the Physicist*. New York: AIP Press, 1995.

Acknowledgments

I'd like to thank Finn Aaserud for his generosity; my editors, Wendy Wolf and Hilary Redmon, for their thoughtfulness; my agent, Flip Brophy, for her truthfulness; the staff at the Niels Bohr Library at the American Institute of Physics for their efficiency; and Deborah Rudacille for her indomitability.

Index

P.O. 0005530877 20240829